Foundations of Engineering Mechanics

Series editors

V.I. Babitsky, Loughborough, Leicestershire, UK
Jens Wittenburg, Karlsruhe, Germany

More information about this series at http://www.springer.com/series/3582

Ranjan Ganguli · Vijay Panchore

The Rotating Beam Problem in Helicopter Dynamics

Springer

Ranjan Ganguli
Department of Aerospace Engineering
Indian Institute of Science
Bangalore, Karnataka
India

Vijay Panchore
Department of Aerospace Engineering
Indian Institute of Science
Bangalore, Karnataka
India

ISSN 1612-1384 ISSN 1860-6237 (electronic)
Foundations of Engineering Mechanics
ISBN 978-981-10-6097-7 ISBN 978-981-10-6098-4 (eBook)
https://doi.org/10.1007/978-981-10-6098-4

Library of Congress Control Number: 2017953797

© Springer Nature Singapore Pte Ltd. 2018
This work is subject to copyright. All rights are reserved by the Publisher, whether the whole or part of the material is concerned, specifically the rights of translation, reprinting, reuse of illustrations, recitation, broadcasting, reproduction on microfilms or in any other physical way, and transmission or information storage and retrieval, electronic adaptation, computer software, or by similar or dissimilar methodology now known or hereafter developed.
The use of general descriptive names, registered names, trademarks, service marks, etc. in this publication does not imply, even in the absence of a specific statement, that such names are exempt from the relevant protective laws and regulations and therefore free for general use.
The publisher, the authors and the editors are safe to assume that the advice and information in this book are believed to be true and accurate at the date of publication. Neither the publisher nor the authors or the editors give a warranty, express or implied, with respect to the material contained herein or for any errors or omissions that may have been made. The publisher remains neutral with regard to jurisdictional claims in published maps and institutional affiliations.

Printed on acid-free paper

This Springer imprint is published by Springer Nature
The registered company is Springer Nature Singapore Pte Ltd.
The registered company address is: 152 Beach Road, #21-01/04 Gateway East, Singapore 189721, Singapore

Preface

The helicopter is a complicated dynamic system. A key aspect of the helicopter is the main rotor which provides lift, propulsive thrust, and control capacity. Modeling of helicopter rotor dynamics is complicated by highly flexible blades and aerodynamics. Comprehensive books on helicopter dynamics have been written. However, they typically derive and discuss the rigid blade equations and provide a cursory treatment to the elastic blade equations, aerodynamic modeling, and solution method for blade response and stability. Thus, a large pedagogical gap exists between the helicopter dynamic books and the theory manuals of the comprehensive rotor codes. This book tries to fill this gap.

We explain the basics of helicopter dynamics which are needed for solving helicopter problems. We take a detailed approach toward the problem of elastic rotating blade, a problem which involves a partial differential equation in space and time and requires a numerical solution. Finite element method plays an important role here, first in space and later in time.

Chapter 1 provides a background on the vibration of discrete and continuous system. It introduces the momentum theory and blade element theory and presents the rigid and elastic flapping blade. The elastic blade equation is derived at the end of Chap. 1. This is a partial differential equation with periodic coefficients and forcing terms. Chapter 2 discusses the finite element method and its use to discretize the rotor blade equation in the spatial domain. Methods to calculate the rotating beam natural frequencies and mode shapes are presented. The aerodynamic forces are also formulated on this chapter. Chapter 3 discusses the finite element method in time, a technique which is ideal for calculating the steady response for periodic systems. This method is illustrated for some periodic differential equations and then used for elastic rotor blade problem. A p-version of the finite element in time is introduced in this chapter. Chapter 4 presents the stability analysis of the elastic blade equation. The constant coefficient approach is illustrated for the rotor in hover. Then, Floquet theory is developed for periodic systems in forward flight.

Finally, the last chapter gives some numerical results for a typical helicopter rotor blade. Frequency analysis, blade response calculation, and stability analysis are presented.

This book will help graduate students and researchers to understand the typical derivations and solution methods used in aeroelastic analysis of helicopter blades.

Bangalore, India
2017

Ranjan Ganguli
Vijay Panchore

Contents

1 Introduction 1
   1.1   Free Vibration of a Single-Degree-of-Freedom System 2
   1.2   Free Vibration of a Damped, Single-Degree-of-Freedom System 3
   1.3   Forced Vibration of a Single-Degree-of-Freedom System 5
   1.4   Forced Vibration of a Damped, Single-Degree-of-Freedom System 7
   1.5   Two-Degrees-of-Freedom System 8
   1.6   Free Vibration of a Continuous System 11
   1.7   Hamilton's Principle 18
   1.8   Diagonalization of a Symmetric Matrix 20
   1.9   Transformation of Coordinates 22
   1.10 Momentum Theory for Axial Flight 23
   1.11 Momentum Theory for Forward Flight 25
   1.12 Newton–Raphson Method 28
   1.13 Blade Element Theory 28
   1.14 Derivation of Equation of Motion of Flapping Rigid Blade 31
   1.15 Derivation of Elastic Rotor Blade Equation 36

2 Finite Element Analysis in Space 41
   2.1   Introduction 41
   2.2   Finite Element in Space 41
   2.3   Strong Form of the Equation 41
   2.4   Weak Form of the Equation 42
   2.5   Galerkin's Method 43
   2.6   Shape Function in 1 Dimension 43
   2.7   Shape Function Formulation for Beam Element 44
   2.8   Properties of Shape Function in 1D 46
   2.9   Finite Element Formulation of Rotating Beam 48
   2.10 Centrifugal Force 51

2.11	Shape Function Formulation for Two Elements	52
2.12	FEM Formulation of Rotating Beam with Only One Shape Function (for Free Vibration)	54
2.13	Calculation of Mode Shapes and Frequencies	57
2.14	FEM Formulation of Aerodynamic Force for Rotor Problem	57

3 Finite Element in Time 61
   3.1 Introduction 61
   3.2 Selection of Shape Function in Time 62
   3.3 Finite Element in Time Example 64
   3.4 Solution of Coupled Differential Equations with Finite Element in Time 68
   3.5 Enforcing Periodicity in the System 70
   3.6 Advantage of Choosing an Element from (0 to 2π), p-Version of Finite Element in Time 71
   3.7 Selection of Number of Nodes 72
   3.8 Effect of Forcing Term in Finite Element in Time ... 72
   3.9 Finite Difference Method (Runge–Kutta Fourth Order) ... 77

4 Stability Analysis 83
   4.1 Introduction 83
   4.2 Stability Analysis of Equations with Constant Coefficients ... 83
   4.3 Stability Analysis of a Coupled Differential Equations with Constant Coefficients 84
   4.4 Stability Analysis of the Equation with Periodic Coefficients, Floquet Theory 87
   4.5 Analytical Solution with the Floquet Theory 88
   4.6 Numerical Method to Evaluate a Transition Matrix ... 89
   4.7 Stability Analysis for Rotor Problem 90

5 Helicopter Rotor Results 93
   5.1 Inputs 93
   5.2 Result 1 (Mode Shapes and Frequencies of the Rotating Beam) 93
   5.3 Result 2 (Response of the Rotor Blade with the Uniform Inflow Model, Using Three Different Cases) 95
   5.4 Result 3 (Response of the Rotor Blade with the Linear Inflow Model, Using Three Different Cases) 95
   5.5 Result 4 (Stability in Hover Condition) 96
   5.6 Result 5 (Stability in Forward Flight) 97
   5.7 Summary and Conclusions 97

References 99

About the Authors

Prof. Ranjan Ganguli obtained his M.S. and Ph.D. in Aerospace Engineering from the University of Maryland, College Park, in 1991 and 1994, respectively, and his B.Tech. in Aerospace Engineering from the Indian Institute of Technology, Kharagpur, in 1989. Following his Ph.D., he worked at the Alfred Gessow Rotorcraft Center of the University of Maryland as Assistant Research Scientist until 1997 on projects on rotorcraft health monitoring and vibratory load validation for the Naval Surface Warfare Center and United Technology Research Center, respectively. He also worked at the GE Research Laboratory in Schenectady, New York, and at Pratt and Whitney, East Hartford, Connecticut, from 1997 to 2000. He joined the Aerospace Engineering Department of the Indian Institute of Science, Bangalore, as Assistant Professor in July 2000. He was promoted to Associate Professor in 2005 and to Full Professor in 2009. He is currently the Satish Dhawan Chair Professor at the Indian Institute of Science, Bangalore. He has held visiting positions at TU Braunschweig, University of Ulm, and Max Planck Institute of Metal Research, Stuttgart, Germany; University Paul Sabatier and Institute of Mathematics, Toulouse, France; Konkuk University, South Korea; the University of Michigan, Ann Arbor, USA; and the Nanyang Technological University, Singapore. Professor Ganguli's research interests are in helicopter aeromechanics, aeroelasticity, structural dynamics, composite and smart structures, design optimization, finite element methods, and health monitoring. He has published 178 articles in refereed journals and over 100 conference papers. He received the American Society of Mechanical Engineers (ASME) best paper award in 2001, the Golden Jubilee award of the Aeronautical Society of India in 2002, the Alexander von Humboldt fellowship in 2007, and the Fulbright Senior Research fellowship in 2010. Professor Ganguli is a Fellow of ASME; a Fellow of the Royal Aeronautical Society, UK; a Fellow of the Indian National Academy of Engineering; a Fellow of the Aeronautical Society of India; an Associate Fellow of the American Institute of Aeronautics and Astronautics; and a Senior Member of the Institute of Electrical and Electronics Engineers (IEEE). He has taught courses on flight and space mechanics, engineering optimization, helicopter dynamics, aircraft structures, structural mechanics, aeroelasticity, and navigation. He has supervised the thesis of

15 Ph.D. and 35 postgraduate students. He has written books on "Engineering Optimization" and "Gas Turbine Diagnostics," both published by CRC Press, New York, and books titled "Structural Damage Detection using Genetic Fuzzy Systems" and "Smart Helicopter Rotors," published by Springer.

Vijay Panchore obtained his B.Tech. in Industrial Engineering from the National Institute of Technology, Jalandhar, in 2009, M.Des. from the Indian Institute of Science, Bangalore, in 2011, and Ph.D. in Aerospace Engineering from the Indian Institute of Science, Bangalore, in 2017. He works in the area of finite element method, meshless methods, and helicopter dynamics. He has published two international journal papers, where finite element method and meshless methods were used to solve the rotating beam problems.

List of Figures

Fig. 1.1	Spring–mass system.	2
Fig. 1.2	Displacement-timegraph (spring–mass system)	3
Fig. 1.3	Damped vibration system.	3
Fig. 1.4	Graphical representation of an over-damped system (*a*), a critically damped system (*b*), and an under-damped system (*c*)	5
Fig. 1.5	Forced vibration of the spring–mass system	6
Fig. 1.6	Forced vibration of the mass–spring–damper system	7
Fig. 1.7	MATLAB plot of "dynamic amplification factor versus frequency ratio".	9
Fig. 1.8	**a** Two-degrees-of-freedom system. **b** Free-body diagram of two-degrees-of-freedom spring–mass–damper system	9
Fig. 1.9	Simply-supported beam with boundary conditions.	12
Fig. 1.10	**a** First mode shape of simply-supported beam. **b** Second mode shape of simply-supported beam. **c** Third mode shape of simply-supported beam	14
Fig. 1.11	Fixed-free beam with boundary conditions	14
Fig. 1.12	**a** First mode shape of fixed-free beam. **b** Second mode shape of fixed-free beam. **c** Third mode shape of fixed-free beam	16
Fig. 1.13	Simply-supported-free beam with boundary conditions	17
Fig. 1.14	**a** First mode shape of simply-supported-free beam. **b** Second mode shape of simply-supported-free beam. **c** Third mode shape of simply-supported-free beam.	19
Fig. 1.15	Simple pendulum.	19
Fig. 1.16	Air flow through the control volume in momentum theory for axial flight.	23
Fig. 1.17	Glauert flow model for momentum analysis of a rotor in forward flight.	26
Fig. 1.18	**a** Blade element theory. **b** Blade element theory. **c** Blade element theory.	29

Fig. 1.19	Tangential and perpendicular components of the flow velocity	30
Fig. 1.20	Forces acting on a small element of a rigid rotor blade	31
Fig. 1.21	**a** Deflection of an elastic rotor blade. **b** Force diagram	36
Fig. 2.1	Elastic bar subjected to uniform load	42
Fig. 2.2	Bar element for shape function formulation	44
Fig. 2.3	Beam element for shape function formulation	44
Fig. 2.4	Two elements in a bar for FEM in space	47
Fig. 2.5	Bar element (shape function properties)	48
Fig. 2.6	Centrifugal force on the rotating beam	52
Fig. 2.7	Shape function formulation of two elements	53
Fig. 2.8	FEM formulation of rotating beam with the shape function of one element (for free vibration)	54
Fig. 3.1	**a** Finite element in time with periodic conditions (displacement), **b** finite element in time with periodic conditions (velocity), **c** finite element in time with initial conditions (displacement), **d** finite element in time with initial conditions (velocity)	67
Fig. 3.2	Finite element in time for coupled differential equations	71
Fig. 3.3	**a** Selection of number of nodes (1 element, 6 nodes), **b** selection of number of nodes (1 element, 11 nodes), **c** selection of number of nodes (1 element, 17 nodes)	73
Fig. 3.4	**a** Effect of forcing (1 element, 11 nodes, $f(\psi) = \sin(\psi) + \cos(\psi)$), **b** effect of forcing (1 element, 11 nodes, $f(\psi) = \sin(\psi) + \cos(\psi) + \sin(2\psi) + \cos(2\psi)$), **c** effect of forcing (1 element, 11 nodes, $f(\psi) = \sin(\psi) + \cos(\psi) + \sin(2\psi) + \cos(2\psi) + \sin(3\psi) + \cos(3\psi)$), **d** effect of forcing (1 element, 11 nodes, $f(\psi) = \sin(\psi) + \cos(\psi) + \sin(2\psi) + \cos(2\psi) + \sin(3\psi) + \cos(3\psi)$)	76
Fig. 3.5	Runge–Kutta fourth-order result	80
Fig. 4.1	**a** Stable system, root locus plot of differential Eq. (4.3), **b** unstable system, root locus plot of differential Eq. (4.4), **c** stable system, solution of differential Eq. (4.3), **d** unstable system, solution of differential Eq. (4.4)	85
Fig. 5.1	Campbell diagram	94
Fig. 5.2	Response with uniform inflow	95
Fig. 5.3	Response with linear inflow	96
Fig. 5.4	Stability in hover condition with varying Lock number	96
Fig. 5.5	Stability in forward flight condition	97

List of Tables

Table 5.1 Inputs for the elastic rotor problem. 94
Table 5.2 Non-dimensional rotating frequencies 94

Chapter 1
Introduction

Helicopters are important flight vehicles which are prone to high vibration caused by an unsteady aerodynamic environment and highly flexible rotating blades. Such high vibrations can result in damage to structural and avionics components, passenger discomfort, and high maintenance costs. The prediction of vibration is therefore a major problem in helicopter engineering. The rotating beam is the fundamental mathematical model for the helicopter rotor blade. Rotor aeroelastic analysis codes typically use rotating beam models, and knowledge of these structures is very necessary for the helicopter dynamics researchers. Typically, calculation of rotating frequencies, blade response, and aeroelastic stability are the key components of helicopter aeroelastic analysis. Thus, numerical methods for solving these problems are a key tool. Specifically, the rotating beam equation is an important model for helicopter dynamics and is a major pedagogical tool. In this chapter, background material for the development of the rotating beam equation is presented. Vibration terminology which is used in this book is explained, and the aerodynamic forces acting on the rotor blade are investigated. The rotor blade equations for a flapping blade and the equation for a rotating beam are then derived. These equations are key to a sound understanding of helicopter rotor dynamics.

The governing partial differential Eq. (1.1) of an elastic rotor blade in flap bending does not have an analytical solution. The numerical solution involves discretization in space and time. We will spend a large part of this book in deriving, analyzing, and solving this equation:

$$\frac{\partial^2}{\partial x^2}\left(EI\frac{\partial^2 w}{\partial x^2}\right) + m\frac{\partial^2 w}{\partial t^2} + \Omega^2\left[mx\frac{\partial w}{\partial x} - \frac{\partial^2 w}{\partial x^2}\int_x^R mx\,dx\right] = F(x,t) \qquad (1.1)$$

Here, EI is the flexural stiffness of the blade, m is the mass per unit length, Ω is the rotating speed of the blade, w is the flap bending deflection, F is the external force, R is the blade radius, and x and t are the spatial and time coordinates.

To understand the material in this book, a good understanding of structural dynamics and helicopter aerodynamics is required. From Sects. 1.1–1.9, we review the basics of structural dynamics. Since the elastic rotor blade is a continuous system, we start with a single-degree-of-freedom system and end with a continuous system. The basics of helicopter aerodynamics including momentum theory and blade element theory are covered later in this chapter. We also discuss the rigid rotor blade flap equation and end this chapter with the derivation of the elastic rotor blade flap equation, including the forcing term on the right-hand side.

1.1 Free Vibration of a Single-Degree-of-Freedom System

Consider Fig. 1.1, which is a spring–mass system. The mass m is suspended by a spring of stiffness k. The mass is then pulled by a distance x and released. The mass then vibrates about an equilibrium position.

The governing differential equation is given by

$$m\ddot{x} + kx = 0 \tag{1.2}$$

Assume the solution $x = C_1 \cos(\omega_n t) + C_2 \sin(\omega_n t)$
At $t = 0, x(0) = C_1, \dot{x}(0)/\omega_n = C_2$
Substituting C_1 and C_2 in the assumed solution, we get the system response as

$$x = x(0)\cos(\omega_n t) + \frac{\dot{x}(0)}{\omega_n}\sin(\omega_n t) \tag{1.3}$$

where $\omega_n = \sqrt{\frac{k}{m}}$ rad/s is the natural frequency of the spring–mass system. The response of the spring–mass system is given in Fig. 1.2.

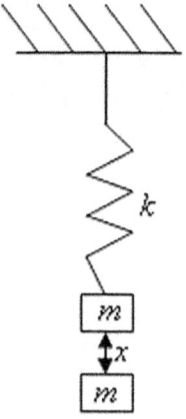

Fig. 1.1 Spring–mass system

1.1 Free Vibration of a Single-Degree-of-Freedom System

Fig. 1.2 Displacement-time graph (spring–mass system)

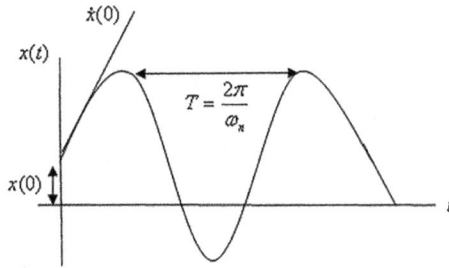

We can write Eq. (1.3) as

$$x = C\sin(\omega_n t + \phi), \qquad (1.4)$$

where $C = \sqrt{x(0)^2 + \frac{\dot{x}(0)^2}{\omega_n^2}}$ and $\phi = \tan^{-1}\left(\frac{x(0)}{\dot{x}(0)\omega_n}\right)$ are the amplitude and phase of the response.

1.2 Free Vibration of a Damped, Single-Degree-of-Freedom System

Consider Fig. 1.3, which shows the damped vibration case. Here, c represents a viscous damper.

Governing equation for the spring–mass–damper system is given by

$$m\ddot{x} + c\dot{x} + kx = 0 \qquad (1.5)$$

Assume the solution $x = Ae^{st}$ and substitute it into Eq. (1.5) yields

$$(ms^2 + cs + k)Ae^{st} = 0 \qquad (1.6)$$

Fig. 1.3 Damped vibration system

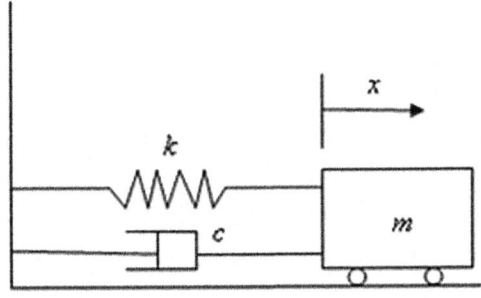

or

$$s = \frac{-c \pm \sqrt{c^2 - 4mk}}{2m} \Rightarrow s = \frac{-c}{2m} \pm \sqrt{\left(\frac{c}{2m}\right)^2 - \frac{k}{m}}$$

or

$$s = \frac{-c}{2m} \pm \sqrt{\left(\frac{c}{2m}\right)^2 - \omega_n^2} \tag{1.7}$$

Here, we define the damping ratio $\zeta = \frac{c}{c_c} = \frac{c}{2m\omega_n}$, where c_c is the critical damping.

We write Eq. (1.7) as

$$s = -\zeta\omega_n \pm \omega_n\sqrt{\zeta^2 - 1} \tag{1.8}$$

There are three possible situations depending on the value of ζ.

If $\left(\frac{c}{2m}\right)^2 - \omega_n^2 > 0$, we get two real roots. In more compact form, if $\zeta > 1$, we get two real roots. This is an over-damped system.

If $\left(\frac{c}{2m}\right)^2 - \omega_n^2 < 0$, we get two complex roots. In more compact form, if $\zeta < 1$, we get two complex roots. This is an under-damped system.

If $\left(\frac{c}{2m}\right)^2 - \omega_n^2 = 0$, we get one repeated real root. In more compact form, if $\zeta = 1$, we get one repeated real root. This is a critically damped system.

For an over-damped case, solution is given by

$$x = C_1 e^{\left(-\zeta\omega_n + \omega_n\sqrt{\zeta^2-1}\right)t} + C_2 e^{\left(-\zeta\omega_n - \omega_n\sqrt{\zeta^2-1}\right)t}$$

or

$$x = e^{-\zeta\omega_n t}\left(C_1 e^{\omega_n t\sqrt{\zeta^2-1}} + C_2 e^{-\omega_n t\sqrt{\zeta^2-1}}\right). \tag{1.9}$$

This is also called the system response of the damped system.

For an under-damped case, solution is given by

$$x = e^{-\zeta\omega_n t}\left(C_1 e^{i\omega_n t\sqrt{1-\zeta^2}} + C_2 e^{-i\omega_n t\sqrt{1-\zeta^2}}\right)$$

or

$$x = e^{-\zeta\omega_n t}\left(C_1 \sin(\omega_n\sqrt{1-\zeta^2}\,t) + C_2 \cos(\omega_n\sqrt{1-\zeta^2}\,t)\right) \tag{1.10}$$

1.2 Free Vibration of a Damped, Single-Degree-of-Freedom System

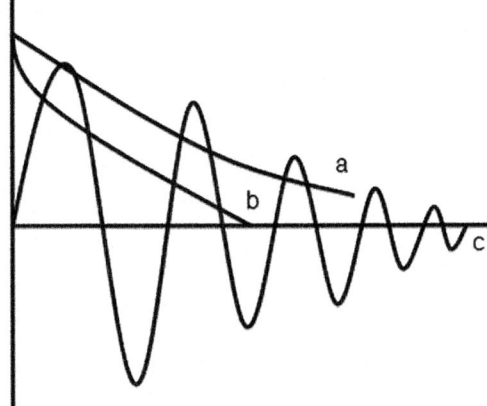

Fig. 1.4 Graphical representation of an over-damped system (*a*), a critically damped system (*b*), and an under-damped system (*c*)

For a critically damped system, solution is given by

$$x = e^{-\zeta \omega_n t}(C_1 + C_2 t) \tag{1.11}$$

Consider Fig. 1.4, where curve (a) shows an over-damped system, curve (b) shows a critically damped system, and curve (c) shows an under-damped system. Damping ensures that the system response dissipates after some time. Adding damping to a system is useful for reducing vibration.

1.3 Forced Vibration of a Single-Degree-of-Freedom System

Consider Fig. 1.5, which shows the forced vibration case. Here, an external force $f(t)$ is applied to the mass-spring system.

The governing equation is given by

$$m\ddot{x} + kx = f(t)$$

Here, $f(t) = P_0 \sin(\bar{\omega} t)$, P_0 is the amplitude of the sinusoidal forcing and $\bar{\omega}$ is its frequency.

We rewrite the governing equation as

$$m\ddot{x} + kx = P_0 \sin(\bar{\omega} t) \tag{1.12}$$

Fig. 1.5 Forced vibration of the spring–mass system

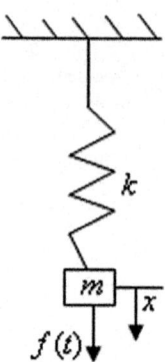

Solution to Eq. (1.12) is given by

$$x = x_h + x_p$$

where x_h is the homogenous solution and x_p is the particular solution.

The homogenous solution is identical to the solution to the system with no forcing term:

$$x_h = C_1 \cos(\omega_n t) + C_2 \sin(\omega_n t) \qquad (1.13)$$

For the particular solution (x_p), we assume the solution to be

$$x_p = A_1 \cos(\bar{\omega} t) + A_2 \sin(\bar{\omega} t) \qquad (1.14)$$

Substituting x_p in Eq. (1.12), we get $A_1 = 0$ and $A_2 = \frac{\frac{P_0}{k}}{1 - \frac{\bar{\omega}^2}{\omega_n^2}}$.

The complete solution is given by

$$x = x_h + x_p$$

or

$$x = C_1 \cos(\omega_n t) + C_2 \sin(\omega_n t) + \frac{P_0}{k\left(1 - \frac{\bar{\omega}^2}{\omega_n^2}\right)} \sin(\bar{\omega} t) \qquad (1.15)$$

or

$$x = C_1 \cos(\omega_n t) + C_2 \sin(\omega_n t) + \frac{P_0}{k(1 - \beta^2)} \sin(\bar{\omega} t) \qquad (1.16)$$

where $\beta = \frac{\bar{\omega}}{\omega_n}$ (frequency ratio). Recall that ω_n is the natural frequency of the spring–mass system and $\bar{\omega}$ is the forcing frequency. The natural frequency of the system is its intrinsic property which depends on stiffness (k) and mass (m) only. From Eq. (1.16), we see that the response becomes infinite when $\beta = 1$ or $\bar{\omega} = \omega_n$. This condition is known as resonance.

1.4 Forced Vibration of a Damped, Single-Degree-of-Freedom System

Consider Fig. 1.6, which shows the forced vibration of the mass–spring–damper system. Here, an external force $f(t) = P_0 \sin \bar{\omega} t$ is applied to the system.

The governing equation is given by

$$m\ddot{x} + c\dot{x} + kx = P_O \sin(\bar{\omega} t) \tag{1.17}$$

The solution to Eq. (1.17) is given by

$$x = x_h + x_p$$

where x_h is the homogenous solution and x_p is the particular solution.

The homogenous solution was derived in Sect. 1.2.

In a real-life situation, we typically deal with an under-damped vibration. Therefore, we take an under-damped vibration case in this section:

$$x_h = e^{-\zeta \omega_n t} \left(C_1 \sin(\omega_n \sqrt{1 - \zeta^2} t) + C_2 \cos(\omega_n \sqrt{1 - \zeta^2} t) \right) \tag{1.18}$$

For the particular solution (x_p), we assume the solution to be

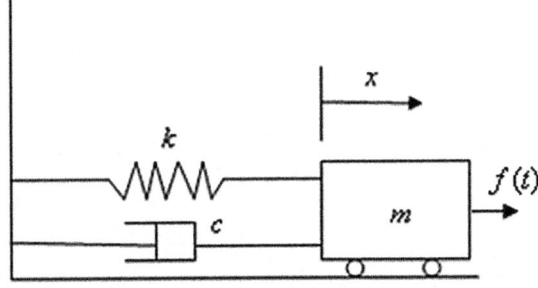

Fig. 1.6 Forced vibration of the mass–spring–damper system

$$x_p = A_1 \cos(\bar{\omega}t) + A_2 \sin(\bar{\omega}t) \tag{1.19}$$

Substituting x_p in Eq. (1.17), we get

$$A_2 = \frac{P_0}{k}\left(\frac{1-\beta^2}{(1-\beta^2)^2 + (2\zeta\beta)^2}\right) \text{ and } A_1 = -\frac{P_0}{k}\left(\frac{2\zeta\beta}{(1-\beta^2)^2 + (2\zeta\beta)^2}\right).$$

The complete solution is given by

$$x = x_h + x_p$$

or

$$\begin{aligned} x = e^{-\zeta\omega_n t}\left(C_1 \sin(\omega_n\sqrt{1-\zeta^2}t) + C_2 \cos(\omega_n\sqrt{1-\zeta^2}t)\right) \\ + \frac{P_0}{k\left((1-\beta^2)^2 + (2\zeta\beta)^2\right)}\left((1-\beta^2)\sin(\bar{\omega}t) - 2\zeta\beta\cos(\bar{\omega}t)\right) \end{aligned} \tag{1.20}$$

$$x = e^{-\zeta\omega_n t}\left(C_1 \sin(\omega_n\sqrt{1-\zeta^2}t) + C_2 \cos\left(\omega_n\sqrt{1-\zeta^2}t\right)\right) + D\frac{P_0}{k}\sin(\bar{\omega}t - \varphi) \tag{1.21}$$

where $D = \frac{1}{\left((1-\beta^2)^2 + (2\zeta\beta)^2\right)^{1/2}}$ and $\phi = \tan^{-1}\left(\frac{2\zeta\beta}{1-\beta^2}\right)$.

Here, D is the dynamic amplification factor. We can see that when $\zeta \to 0$ and $\beta \to 1$, $D \to \infty$, a condition known as resonance where the amplitude of the system increases in an unbounded manner.

Figure 1.7 shows how the dynamic amplification factor varies with the frequency ratio.

We see in Fig. 1.7, as the damping factor increases, the amplitude of the dynamic amplification factor decreases. Adding damping to a system results in amelioration of the resonance peak. Also, it is advisable to keep the frequency ratio away from one. Considerable effort is spent in the development of dampers and for frequency placement for vibration reduction.

1.5 Two-Degrees-of-Freedom System

Consider Fig. 1.8a, a two-degrees-of-freedom system. There are two masses m_1 and m_2, two springs k_1 and k_2, and two dampers c_1 and c_2. A force P_1 is applied to mass m_1, and a force P_2 is applied to mass m_2. The degrees of freedom are the coordinates

1.5 Two-Degrees-of-Freedom System

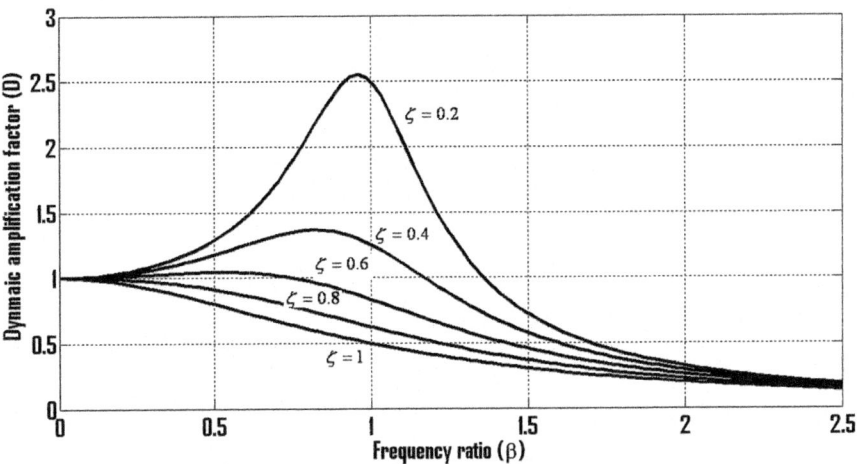

Fig. 1.7 MATLAB plot of "dynamic amplification factor versus frequency ratio"

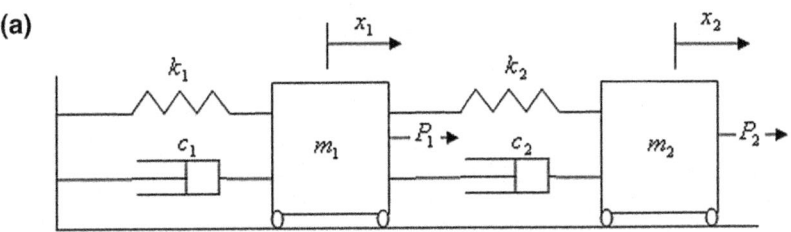

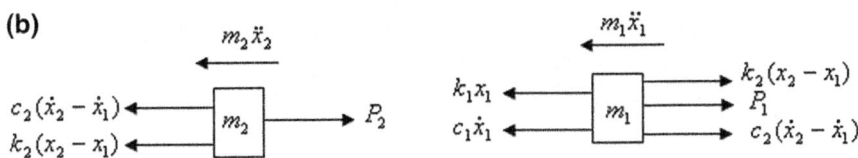

Fig. 1.8 a Two-degrees-of-freedom system. **b** Free-body diagram of two-degrees-of-freedom spring–mass–damper system

x_1 and x_2 which represent the motion of the two masses. In Fig. 1.8b, we draw a free-body diagram for each mass and then write the equations of motion.

We write the governing equation as

$$\begin{bmatrix} m_1 & 0 \\ 0 & m_2 \end{bmatrix} \begin{bmatrix} \ddot{x}_1 \\ \ddot{x}_2 \end{bmatrix} + \begin{bmatrix} c_1 + c_2 & -c_2 \\ -c_2 & c_2 \end{bmatrix} \begin{bmatrix} \dot{x}_1 \\ \dot{x}_2 \end{bmatrix} + \begin{bmatrix} k_1 + k_2 & -k_2 \\ -k_2 & k_2 \end{bmatrix} \begin{bmatrix} x_1 \\ x_2 \end{bmatrix} = \begin{bmatrix} P_1 \\ P_2 \end{bmatrix}. \tag{1.22}$$

We write the free vibration equation by ignoring the damping term and the force vector

$$\begin{bmatrix} m_1 & 0 \\ 0 & m_2 \end{bmatrix} \begin{bmatrix} \ddot{x}_1 \\ \ddot{x}_2 \end{bmatrix} + \begin{bmatrix} k_1 + k_2 & -k_2 \\ -k_2 & k_2 \end{bmatrix} \begin{bmatrix} x_1 \\ x_2 \end{bmatrix} = \begin{bmatrix} 0 \\ 0 \end{bmatrix}. \quad (1.23)$$

Assume the solution

$$\begin{bmatrix} x_1 \\ x_2 \end{bmatrix} = \begin{bmatrix} C_1 \sin(\omega t - \phi) \\ C_2 \sin(\omega t - \phi) \end{bmatrix}.$$

Substituting the assumed solution into Eq. (1.23), we get

$$\begin{bmatrix} -m_1 \omega^2 + k_1 + k_2 & -k_2 \\ -k_2 & -m_2 \omega^2 + k_2 \end{bmatrix} \begin{bmatrix} C_1 \sin(\omega t - \phi) \\ C_2 \sin(\omega t - \phi) \end{bmatrix} = \begin{bmatrix} 0 \\ 0 \end{bmatrix} \quad (1.24)$$

For a non-trivial solution

$$\begin{vmatrix} -m_1 \omega^2 + k_1 + k_2 & -k_2 \\ -k_2 & -m_2 \omega^2 + k_2 \end{vmatrix} = 0$$

or

$$\omega^4 - \left(\frac{k_1 + k_2}{m_1} + \frac{k_2}{m_2} \right) \omega^2 + \frac{k_1 k_2}{m_1 m_2} = 0$$

or

$$\lambda = \omega^2 = \frac{1}{2} \left(\frac{k_1 + k_2}{m_1} + \frac{k_2}{m_2} \right) \pm \frac{1}{2} \left\{ \left(\frac{k_1 + k_2}{m_1} + \frac{k_2}{m_2} \right)^2 - 4 \frac{k_1 k_2}{m_1 m_2} \right\}^{1/2}. \quad (1.25)$$

From Eq. (1.25), we can get two natural frequencies ω_1 and ω_2.
From the homogenous Eq. (1.24), we can get the ratio of C_1 and C_2.

$$r_1 = \frac{C_1^{(1)}}{C_2^{(1)}} = \frac{K_2}{-m_1 \omega_1^2 + K_1 + K_2} = \frac{m_2 \omega_1^2 + K_2}{K_2} \quad (1.26)$$

1.5 Two-Degrees-of-Freedom System

$$r_2 = \frac{C_1^{(2)}}{C_2^{(2)}} = \frac{K_2}{-m_1\omega_2^2 + K_1 + K_2} = \frac{m_2\omega_2^2 + K_2}{K_2}. \qquad (1.27)$$

The two-degrees-of-freedom system has two modes of vibration corresponding to two natural frequencies ω_1 and ω_2, respectively, and can be expressed as

$$[C^{(1)}] = \begin{bmatrix} C_1^{(1)} \\ C_2^{(1)} \end{bmatrix} = \begin{bmatrix} C_1^{(1)} \\ rC_1^{(1)} \end{bmatrix} \qquad (1.28)$$

$$[C^{(2)}] = \begin{bmatrix} C_1^{(2)} \\ C_2^{(2)} \end{bmatrix} = \begin{bmatrix} C_1^{(2)} \\ rC_1^{(2)} \end{bmatrix} \qquad (1.29)$$

where $[C^{(1)}]$ and $[C^{(2)}]$ are the two modes of vibration.

Free vibration response of the two-degrees-of-freedom system is given by

$$\begin{bmatrix} x_1 \\ x_2 \end{bmatrix} = \begin{bmatrix} C_1^{(1)} \\ rC_1^{(1)} \end{bmatrix} \sin(\omega t - \phi) + \begin{bmatrix} C_1^{(2)} \\ rC_1^{(2)} \end{bmatrix} \sin(\omega t - \phi) \qquad (1.30)$$

The concepts and derivation for the two-degrees-of-freedom system can be extended to a multi-degree-of-freedom system.

1.6 Free Vibration of a Continuous System

In a continuous system, we consider continuous distribution of mass, damping, and elasticity. Rods, beams, and plates are typical examples of such systems. Helicopter rotor blades are typically modeled as beams and so we discuss beams as an example of a continuous system.

Governing differential equation of a beam is given by

$$\frac{\partial^2}{\partial x^2}\left(EI\frac{\partial^2 w}{\partial x^2}\right) + m\ddot{w} = 0 \qquad (1.31)$$

where EI is the flexural stiffness, m is the mass per unit length, and w is the deflection in the transverse direction.

For constant EI, we get

$$EI\frac{\partial^4 w}{\partial x^4} + m\ddot{w} = 0 \qquad (1.32)$$

Assume the solution $w(x,t) = \bar{w}(x)e^{i\omega t}$, we get

$$EI\frac{d^4\bar{w}(x)}{dx^4}e^{i\omega t} - w^2 m\bar{w}(x)e^{i\omega t} = 0$$

$$EI\frac{d^4\bar{w}(x)}{dx^4} - \omega^2 m\bar{w}(x) = 0 \qquad (1.33)$$

This equation is an ordinary differential equation in space:

$$\frac{d^4\bar{w}(x)}{dx^4} - \frac{m\omega^2}{EI}\bar{w}(x) = 0 \qquad (1.34)$$

Assume the solution $\bar{w}(x) = e^{px}$, we get

$$p^4 e^{px} - \frac{m\omega^2}{EI}e^{px} = 0 \Rightarrow p^4 = \frac{m\omega^2}{EI}$$

or

$$p^4 = \lambda^4$$

where $\lambda = \left(\frac{m\omega^2}{EI}\right)^{1/4}$

$$p = \lambda, -\lambda, i\lambda, -i\lambda$$

The solution is given by

$$\bar{w}(x) = G_1 e^{\lambda x} + G_2 e^{-\lambda x} + G_3 e^{i\lambda x} + G_4 e^{-i\lambda x}$$

$$\bar{w}(x) = C_1 \sinh(\lambda x) + C_2 \cosh(\lambda x) + C_3 \sin(\lambda x) + C_4 \cos(\lambda x) \qquad (1.35)$$

where $G_1, G_2, G_3, G_4, C_1, C_2, C_3,$ and C_4 are constants. Here, we need four boundary conditions to find the deflection.

Consider a simply-supported beam, and boundary conditions are shown in Fig. 1.9. This type of boundary condition gives simple mode shapes and frequencies and is given for illustration.

Here, boundary conditions are

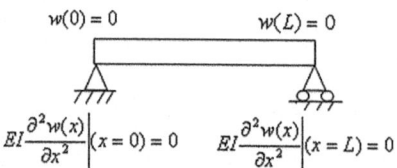

Fig. 1.9 Simply-supported beam with boundary conditions

1.6 Free Vibration of a Continuous System

$$w(0) = 0, \text{EI}\frac{\partial^2 w(x)}{\partial x^2}\bigg|(x=0) = 0, w(L) = 0, \text{ and } \text{EI}\frac{\partial^2 w(x)}{\partial x^2}\bigg|(x=L) = 0.$$

We apply the boundary conditions to Eq. (1.35) and get

$$0 * C_1 + 1 * C_2 + 0 * C_3 + 1 * C_4 = 0$$

$$0 * C_1 + 1 * C_2 + 0 * C_3 - 1 * C_4 = 0$$

$$\sin h(\lambda L) * C_1 + \cos h(\lambda L) * C_2 + \sin(\lambda L) * C_3 + \cos(\lambda L) * C_4 = 0$$

$$\sin h(\lambda L) * C_1 + \cos h(\lambda L) * C_2 - \sin(\lambda L) * C_3 - \cos(\lambda L) * C_4 = 0$$

Solving, we get

$$C_3 \sin(\lambda L) = 0$$

or

$$\lambda L = n\pi \Rightarrow \left(\frac{m\omega^2}{EI}\right)^{1/4} L = n\pi \Rightarrow \omega = n^2\pi^2 \sqrt{\frac{EI}{mL^4}}$$

Substituting $r = n$, we get

$$\omega_r = r^2\pi^2 \sqrt{\frac{EI}{L^4 m}} \qquad (1.36)$$

where $\omega_1, \omega_2 \ldots \omega_r$ are the natural frequencies of the system and the respective mode shapes are given by

$$\bar{w}_r(x) = C_3 \sin\left(\frac{rx}{L}\right) \qquad (1.37)$$

The first three natural frequencies are

$$\omega_1 = \pi^2 \sqrt{\frac{EI}{mL^4}}, \omega_2 = 4\pi^2 \sqrt{\frac{EI}{mL^4}}, \text{ and } \omega_3 = 9\pi^2 \sqrt{\frac{EI}{mL^4}}$$

The first three mode shapes are

$$\bar{w}_1(x) = C_3 \sin\left(\frac{\pi x}{L}\right), \bar{w}_2(x) = C_3 \sin\left(\frac{2\pi x}{L}\right), \text{ and } \bar{w}_3(x) = C_3 \sin\left(\frac{3\pi x}{L}\right).$$

Figure 1.10 shows the plots of the first three mode shapes. The complete solution is given by

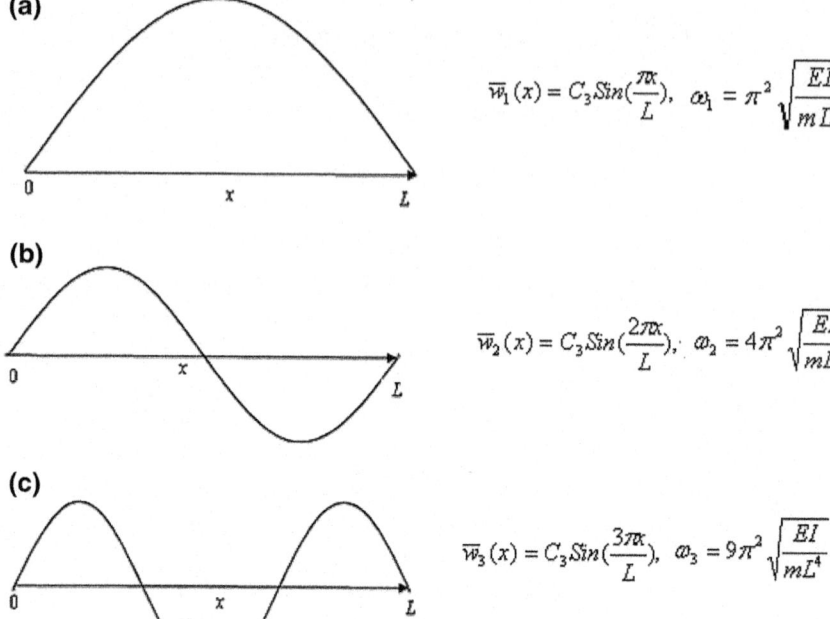

Fig. 1.10 **a** First mode shape of simply-supported beam. **b** Second mode shape of simply-supported beam. **c** Third mode shape of simply-supported beam

Fig. 1.11 Fixed-free beam with boundary conditions

$$w(x,t) = \sin\left(\frac{r\pi x}{L}\right) e^{i\omega t}. \quad (1.38)$$

The rotor blade is typically a fixed-free beam (hingeless rotor) or a simply-supported-free beam (articulated rotor). We discuss these two types of boundary conditions next.

Fixed-free beam

Boundary conditions for a fixed-free beam are (Fig. 1.11)

1.6 Free Vibration of a Continuous System

$$w(0) = 0, \frac{\partial w(x)}{\partial x}\bigg|(x=0) = 0, \text{EI}\frac{\partial^2 w(x)}{\partial x^2}\bigg|(x=L) = 0, \text{ and } \text{EI}\frac{\partial^3 w(x)}{\partial x^3}\bigg|(x=L) = 0.$$

Applying boundary conditions $w(0) = 0$ and $\frac{\partial w(x)}{\partial x}\big|(x=0) = 0$ to Eq. (1.35), we get

$$C_2 + C_4 = 0 \text{ and } C_1 + C_3 = 0.$$

We rewrite assumed solution (1.35) as

$$\bar{w}(x) = C_1\{\sin h(\lambda x) - \sin(\lambda x)\} + C_2\{\cos h(\lambda x) - \cos(\lambda x)\} \quad (1.39)$$

Applying boundary conditions $\text{EI}\frac{\partial^2 w(x)}{\partial x^2}\big|(x=L) = 0$ and $\text{EI}\frac{\partial^3 w(x)}{\partial x^3}\big|(x=L) = 0$ to Eq. (1.39), we get

$$\begin{bmatrix} \sin h(\lambda L) + \sin(\lambda L) & \cos h(\lambda L) + \cos(\lambda L) \\ \cos h(\lambda L) + \cos(\lambda L) & \sin h(\lambda L) - \sin(\lambda L) \end{bmatrix} \begin{bmatrix} C_1 \\ C_2 \end{bmatrix} = \begin{bmatrix} 0 \\ 0 \end{bmatrix}$$

For a non-trivial solution

$$\begin{vmatrix} \sin h(\lambda L) + \sin(\lambda L) & \cos h(\lambda L) + \cos(\lambda L) \\ \cos h(\lambda L) + \cos(\lambda L) & \sin h(\lambda L) - \sin(\lambda L) \end{vmatrix} = 0$$

or

$$1 + \cos h(\lambda L)\cos(\lambda L) = 0 \quad (1.40)$$

We solve Eq. (1.40) numerically to get the first three solutions

$$\lambda L = 1.875104, 4.694091, \text{ and } 7.48547$$

Using boundary condition $\text{EI}\frac{\partial^3 w(x)}{\partial x^3}\big|(x=L) = 0$, we get

$$C_2 = -C_1 \frac{\cos h(\lambda L) + \cos(\lambda L)}{\sin h(\lambda L) - \sin(\lambda L)}$$

We rewrite Eq. (1.39) as

$$\bar{w}(x) = C_1\{[\sin h(\lambda x) - \sin(\lambda x)] - \frac{\cos h(\lambda L) + \cos(\lambda L)}{\sin h(\lambda L) - \sin(\lambda L)}[\cos h(\lambda x) - \cos(\lambda x)]\} \quad (1.41)$$

The first three natural frequencies are given by

$$\omega_1 = (1.8751)^2 \sqrt{\frac{EI}{mL^4}}, \omega_2 = (4.6940)^2 \sqrt{\frac{EI}{mL^4}}, \text{ and } \omega_3 = (7.8547)^2 \sqrt{\frac{EI}{mL^4}}$$

The first three mode shapes are given by

$$\bar{w}_1(x) = C_1 \left\{ \left[\sinh\left(1.8751\frac{x}{L}\right) - \sin\left(1.8751\frac{x}{L}\right) \right] \right.$$
$$\left. - \frac{\cosh(1.8751) + \cos(1.8751)}{\sinh(1.8751) - \sin(1.8751)} \left[\cosh\left(1.8751\frac{x}{L}\right) - \cos\left(1.8751\frac{x}{L}\right) \right] \right\},$$

$$\bar{w}_2(x) = C_1 \left\{ \left[\sinh\left(4.6940\frac{x}{L}\right) - \sin\left(4.6940\frac{x}{L}\right) \right] \right.$$
$$\left. - \frac{\cosh(4.6940) + \cos(4.6940)}{\sinh(4.6940) - \sin(4.6940)} \left[\cosh\left(4.6940\frac{x}{L}\right) - \cos\left(4.6940\frac{x}{L}\right) \right] \right\},$$

and

$$\bar{w}_3(x) = C_1 \left\{ \left[\sinh\left(7.8547\frac{x}{L}\right) - \sin\left(7.8547\frac{x}{L}\right) \right] \right.$$
$$\left. - \frac{\cosh(7.8547) + \cos(7.8547)}{\sinh(7.8547) - \sin(7.8547)} \left[\cosh\left(7.8547\frac{x}{L}\right) - \cos\left(7.8547\frac{x}{L}\right) \right] \right\}$$

Figure 1.12 shows the plots of the first three mode shapes. These mode shapes are also known as beam function for a cantilever beam and are used as basis functions for approximate methods.

Fig. 1.12 a First mode shape of fixed-free beam. **b** Second mode shape of fixed-free beam. **c** Third mode shape of fixed-free beam

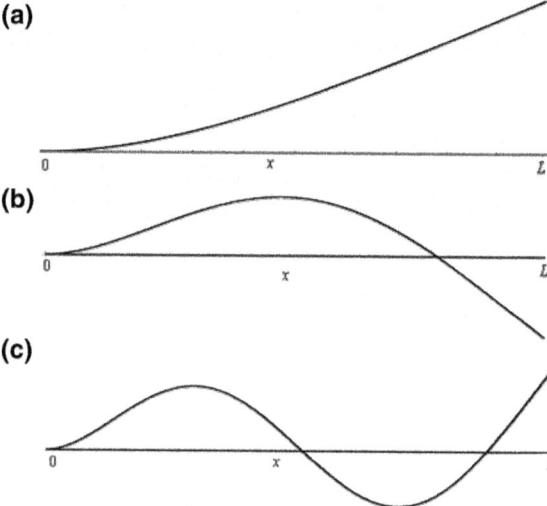

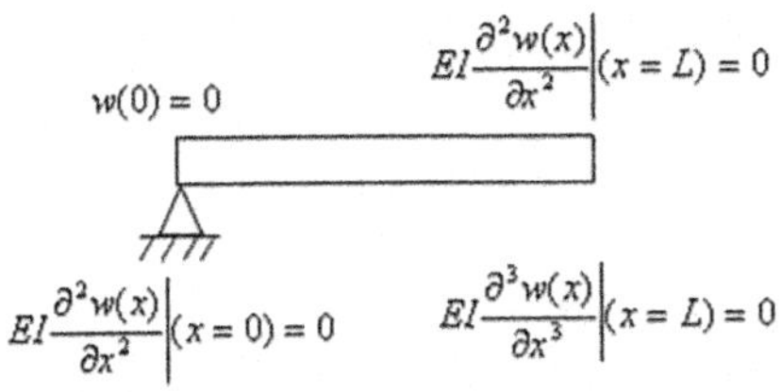

1.6 Free Vibration of a Continuous System

Fig. 1.13 Simply-supported-free beam with boundary conditions

Simply-supported-free beam

Boundary conditions for a simply-supported-free beam are (Fig. 1.13).

$$w(0) = 0, \text{EI}\frac{\partial^2 w(x)}{\partial x^2}\bigg|(x=0) = 0, \text{EI}\frac{\partial^2 w(x)}{\partial x^2}\bigg|(x=L) = 0, \text{ and } \text{EI}\frac{\partial^3 w(x)}{\partial x^3}\bigg|(x=L) = 0.$$

Applying boundary conditions $w(0) = 0$ and $\text{EI}\frac{\partial^2 w(x)}{\partial x^2}\big|(x=0) = 0$ to Eq. (1.35), we get

$$C_4 = 0 \text{ and } C_2 = 0.$$

We rewrite assumed solution (1.35) as

$$\bar{w}(x) = C_1 \sinh(\lambda x) + C_3 \sin(\lambda x) \tag{1.42}$$

Applying boundary conditions $\text{EI}\frac{\partial^2 w(x)}{\partial x^2}\big|(x=L) = 0$ and $\text{EI}\frac{\partial^3 w(x)}{\partial x^3}\big|(x=L) = 0$ to Eq. (1.42), we get

$$\begin{bmatrix} \sinh(\lambda L) & -\sin(\lambda L) \\ \cosh(\lambda L) & -\cos(\lambda L) \end{bmatrix} \begin{bmatrix} C_1 \\ C_3 \end{bmatrix} = \begin{bmatrix} 0 \\ 0 \end{bmatrix}$$

For a non-trivial solution

$$\begin{vmatrix} \sinh(\lambda L) & -\sin(\lambda L) \\ \cosh(\lambda L) & -\cos(\lambda L) \end{vmatrix} = 0 \tag{1.43}$$

We solve Eq. (1.43) numerically to get the first three solutions

$$\lambda L = 3.9266, 7.0685, \text{ and } 10.2101$$

Using boundary condition $EI\frac{\partial^3 w(x)}{\partial x^3}\big|(x = L) = 0$, we get

$$C_3 = C_1 \frac{\cos h(\lambda L)}{\cos(\lambda L)}$$

We rewrite Eq. (1.42) as

$$\bar{w}(x) = C_1 \left\{ \sin h(\lambda x) + \frac{\cos h(\lambda L)}{\cos(\lambda L)} \sin(\lambda x) \right\} \tag{1.44}$$

The first three natural frequencies are given by

$$\omega_1 = (3.9266)^2 \sqrt{\frac{EI}{mL^4}}, \omega_2 = (7.0685)^2 \sqrt{\frac{EI}{mL^4}}, \text{ and } \omega_3 = (10.2101)^2 \sqrt{\frac{EI}{mL^4}}$$

The first three mode shapes are given by

$$\bar{w}_1(x) = C_1 \left\{ \sin h\left(3.9266 \frac{x}{L}\right) + \frac{\cos h(3.9266)}{\cos(3.9266)} \sin\left(3.9266 \frac{x}{L}\right) \right\},$$

$$\bar{w}_2(x) = C_1 \left\{ \sin h\left(7.0685 \frac{x}{L}\right) + \frac{\cos h(7.0685)}{\cos(7.0685)} \sin\left(7.0685 \frac{x}{L}\right) \right\},$$

and

$$\bar{w}_3(x) = C_1 \left\{ \sin h\left(10.2101 \frac{x}{L}\right) + \frac{\cos h(10.2101)}{\cos(10.2101)} \sin\left(10.2101 \frac{x}{L}\right) \right\}$$

Figure 1.10 shows the plots of the first three mode shapes. These are elastic mode shapes (Fig. 1.14).

1.7 Hamilton's Principle

Hamilton's principle is often used to derive the equation of motion for helicopter rotor blades undergoing motion in several directions such as flap and lag bending, axial, and torsion. It states that of all of the possible paths of a mechanical system, the path actually followed is the one that minimizes the time integral of the difference between the kinetic and potential energies. That is, the actual path chosen by the system is the one that makes the variation of the following integral vanish:

$$\delta \int (T - U) dt = 0 \tag{1.45}$$

1.7 Hamilton's Principle

Fig. 1.14 a First mode shape of simply-supported-free beam. **b** Second mode shape of simply-supported-free beam. **c** Third mode shape of simply-supported-free beam

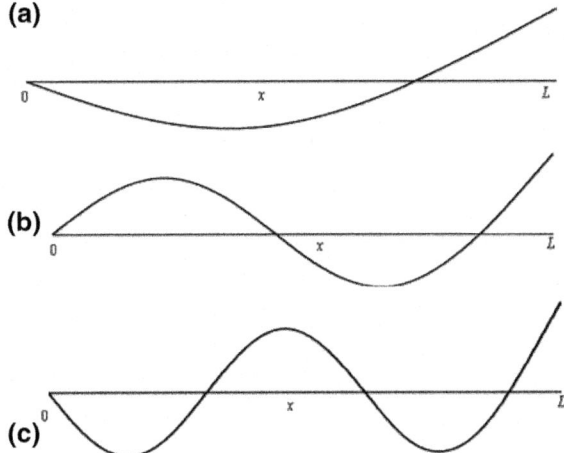

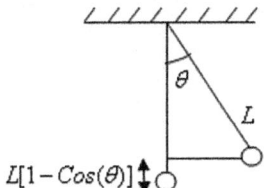

Fig. 1.15 Simple pendulum

where T is the kinetic energy and U is the potential energy.

We can write Eq. (1.45) as

$$\delta \int L dt = 0 \qquad (1.46)$$

where L is the Lagrangian of the system.

The equation of motion is derived using Lagrange's equation

$$\frac{d}{dt}\left(\frac{\partial L}{\partial \dot{q}_i}\right) - \frac{\partial L}{\partial q_i} = 0 \quad (i = 1, 2, 3) \qquad (1.47)$$

Consider Fig. 1.15, the case of a simple pendulum.
Here, $T = \frac{1}{2}mL^2\dot{\theta}^2$ and $U = mgL(1 - \cos(\theta))$

Using the Lagrange's Eq. (1.47), we get the equation of motion of a simple pendulum

$$\ddot{\theta} + \frac{g}{L}\sin(\theta) = 0. \tag{1.48}$$

1.8 Diagonalization of a Symmetric Matrix

In a multi-degree-of-freedom system, the governing equation is given by

$$[M][\ddot{X}] + [C][\dot{X}] + [K][X] = [F(t)] \tag{1.49}$$

Here, $[M]$ and $[K]$ are symmetric matrices.
It would be easier to solve Eq. (1.51) than Eq. (1.50):

$$\begin{bmatrix} m_{11} & 0 & 0 \\ 0 & m_{22} & 0 \\ 0 & 0 & m_{33} \end{bmatrix} \begin{bmatrix} \ddot{x}_1 \\ \ddot{x}_2 \\ \ddot{x}_3 \end{bmatrix} + \begin{bmatrix} k_{11} & & \\ k_{21} & k_{22} & \\ k_{31} & k_{32} & k_{33} \end{bmatrix} \begin{bmatrix} x_1 \\ x_2 \\ x_3 \end{bmatrix} = \begin{bmatrix} f_1(t) \\ f_2(t) \\ f_3(t) \end{bmatrix} \tag{1.50}$$

$$\begin{bmatrix} m_1 & 0 & 0 \\ 0 & m_2 & 0 \\ 0 & 0 & m_3 \end{bmatrix} \begin{bmatrix} \ddot{\zeta}_1 \\ \ddot{\zeta}_2 \\ \ddot{\zeta}_3 \end{bmatrix} + \begin{bmatrix} k_1 & 0 & 0 \\ 0 & k_2 & 0 \\ 0 & 0 & k_3 \end{bmatrix} \begin{bmatrix} \zeta_1 \\ \zeta_2 \\ \zeta_3 \end{bmatrix} = \begin{bmatrix} f_1(t) \\ f_2(t) \\ f_3(t) \end{bmatrix} \tag{1.51}$$

We can diagonalize a symmetric matrix $[A]$ with a matrix $[P]$ such that $[P]^{-1}[A][P]$ is a diagonal matrix.

$$[P]_{nxn} = \begin{bmatrix} [v_1]_{nx1} & [v_2]_{nx1} & [v_3]_{nx1} & \ldots & [v_n]_{nx1} \end{bmatrix} \tag{1.52}$$

where $[v_1]_{nx1}, [v_2]_{nx1} \ldots, \& [v_n]_{nx1}$ are the eigenvectors of the matrix $[A]$.
We know

$$[A]_{nxn}[v_1]_{nx1} = \lambda_1 [v_1]_{nx1} \tag{1.53}$$

or

$$[A]_{nxn}[P]_{nxn} = [P]_{nxn} \begin{bmatrix} \lambda_1 & 0 & 0 & 0 & 0 \\ 0 & \lambda_2 & 0 & 0 & 0 \\ 0 & 0 & . & 0 & 0 \\ 0 & 0 & 0 & . & 0 \\ 0 & 0 & 0 & 0 & \lambda_n \end{bmatrix}_{nxn} \tag{1.54}$$

1.8 Diagonalization of a Symmetric Matrix

or

$$[P]_{nxn}^{-1}[A]_{nxn}[P]_{nxn} = \begin{bmatrix} \lambda_1 & 0 & 0 & 0 & 0 \\ 0 & \lambda_2 & 0 & 0 & 0 \\ 0 & 0 & . & 0 & 0 \\ 0 & 0 & 0 & . & 0 \\ 0 & 0 & 0 & 0 & \lambda_n \end{bmatrix}_{nxn} \quad (1.55)$$

We see that $[P]_{nxn}^{-1}[A]_{nxn}[P]_{nxn}$ is a diagonal matrix.

The eigenvectors of a real symmetric matrix corresponding to distinct eigenvalues are orthogonal.

We rewrite Eq. (1.41)

$$[A]_{nxn}[v_1]_{nx1} = \lambda_1 [v_1]_{nx1}$$

or

$$\left([A]_{nxn}[v_1]_{nx1}\right)^T = \lambda_1 [v_1]_{nx1}^T \Rightarrow [v_1]_{nx1}^T [A]_{nx1}^T = \lambda_1 [v_1]_{nx1}^T$$

or

$$[v_1]_{nx1}^T [A]_{nxn}[v_2]_{nx1} = \lambda_1 [v_1]_{nx1}^T [v_2]_{nx1} \Rightarrow \lambda_2 [v_1]_{nx1}^T [v_2]_{nx1} = \lambda_1 [v_1]_{nx1}^T [v_2]_{nx1}$$

or

$$(\lambda_2 - \lambda_1)[v_1]_{nx1}^T [v_2]_{nx1} = 0 \quad (\lambda_2 \neq \lambda_1)$$

or

$$[v_1]_{nx1}^T [v_2]_{nx1} = 0 \quad (1.56)$$

Equation (1.56) shows that eigenvectors corresponding to distinct eigenvalues are orthogonal.

From Eq. (1.56), we write

$$[P]_{nxn}^T [P]_{nxn} = [I]_{nxn}$$

or

$$[P]_{nxn}^{-1} = [P]_{nxn}^T$$

We rewrite Eq. (1.55) as

$$[P]^T_{nxn}[A]_{nxn}[P]_{nxn} = \begin{bmatrix} \lambda_1 & 0 & 0 & 0 & 0 \\ 0 & \lambda_2 & 0 & 0 & 0 \\ 0 & 0 & . & 0 & 0 \\ 0 & 0 & 0 & . & 0 \\ 0 & 0 & 0 & 0 & \lambda_n \end{bmatrix}_{nxn}. \quad (1.57)$$

1.9 Transformation of Coordinates

We write a multi-degree-of-freedom equation

$$[M][\ddot{X}] + [K][X] = [F(t)] \quad (1.58)$$

Typically, mass and stiffness matrices are fully populated.

We write $[X(t)] = [\phi][\zeta(t)]$, where $[X(t)]$ are the general coordinates, $[\phi]$ is the modal transformation matrix, and $[\zeta(t)]$ is the principal coordinate.

Here, $[\varphi]_{nxn} = \left[[\varphi_1]_{nx1}[\varphi_2]_{nx1}\ldots[\varphi_n]_{nx1}\right]$ and $[\zeta(t)]_{nx1} = [\zeta_1(t)\zeta_2(t)\ldots\zeta_n(t)]^T$

After the coordinate transformation, we get Eq. (1.59):

$$[M][\phi][\ddot{\zeta}(t)] + [K][\phi][\zeta(t)] = [F(t)] \quad (1.59)$$

or

$$[\phi]^T[M][\phi][\ddot{\zeta}(t)] + [\phi]^T[K][\phi][\zeta(t)] = [\phi]^T[F(t)] \quad (1.60)$$

Here,

$$[\phi_i]^T_{nx1}[M]_{nxn}[\phi_j]_{nx1} = \begin{cases} M_i > 0 (i=j) \\ 0 (i \neq j) \end{cases}, [\phi_i]^T_{nx1}[K]_{nxn}[\phi_j]_{nx1} = \begin{cases} \omega_i^2 M_i > 0 (i=j) \\ 0 (i \neq j) \end{cases},$$

or

$$\begin{bmatrix} m_1 & 0 & 0 & 0 & 0 \\ 0 & m_2 & 0 & 0 & 0 \\ 0 & 0 & . & 0 & 0 \\ 0 & 0 & 0 & . & 0 \\ 0 & 0 & 0 & 0 & m_n \end{bmatrix} \begin{bmatrix} \ddot{\zeta}_1(t) \\ \ddot{\zeta}_2(t) \\ . \\ . \\ \ddot{\zeta}_n(t) \end{bmatrix} + \begin{bmatrix} k_1 & 0 & 0 & 0 & 0 \\ 0 & k_2 & 0 & 0 & 0 \\ 0 & 0 & . & 0 & 0 \\ 0 & 0 & 0 & . & 0 \\ 0 & 0 & 0 & 0 & k_n \end{bmatrix} \begin{bmatrix} \zeta_1(t) \\ \zeta_2(t) \\ . \\ . \\ \zeta_n(t) \end{bmatrix} = \begin{bmatrix} f_1(t) \\ f_2(t) \\ . \\ . \\ f_n(t) \end{bmatrix}$$

$$(1.61)$$

Equation (1.61) is a set decoupled equation and is easier to solve than Eq. (1.58). The modal transformation converts a multi-degree-of-freedom system into a series of single-degree-of-freedom systems.

1.10 Momentum Theory for Axial Flight

The forcing function vector for the helicopter rotor blade equation comes from aerodynamics. We review basic helicopter aerodynamics in the next few sections. Momentum theory is a simple approach which can predict rotor inflow and power. In this theory, the helicopter rotor is modeled as an actuator disk. Momentum theory uses the principle of linear momentum conservation and assumes incompressible, irrotational, and steady flow.

Consider Fig. 1.16, where we take a control volume over the plane of the rotor. Here, V_∞ is the flow velocity, V_u and V_l are the induced velocities at upper and lower side of the rotor plane, respectively, V_f is the induced velocity at the slipstream, A_1, A_2, A_3, and A_4 are the areas of the four stations 1, 2, 3, and 4, respectively, and P_1, P_2, P_3, and P_4 are the pressures at the four stations 1, 2, 3, and 4, respectively.

We write the mass conservation equation as

$$\frac{\partial}{\partial t} \iiint_{CV} \rho dv + \iint_{CS} \rho \vec{V}.\vec{n}dA = 0 \qquad (1.62)$$

Since flow is steady $\left(\frac{\partial}{\partial t} \iiint_{CV} \rho dv = 0\right)$, we get the equation

$$\iint_{CS} \rho \vec{V}.\vec{n}dA = 0 \qquad (1.63)$$

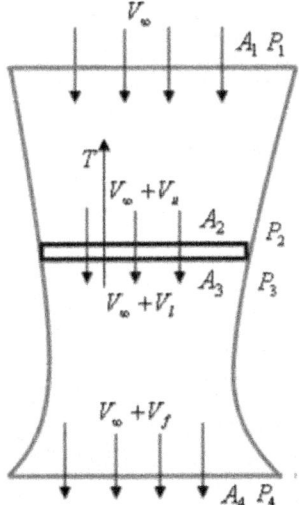

Fig. 1.16 Air flow through the control volume in momentum theory for axial flight

We write Eq. (1.63) for flow from area A_2 to area A_3, to get

$$\rho(V_\infty + V_u)A_2 = \rho(V_\infty + V_l)A_3$$

Since $A_2 = A_3$, we get

$(V_\infty + V_u) = (V_\infty + V_l) \Rightarrow V_u = V_l$, and there is no jump in the velocity across the rotor.

We write the momentum conservation equation as

$$T = \frac{\partial}{\partial t} \iiint_{CV} \vec{V} \rho dv + \iint_{CS} \rho \vec{V}(\vec{V}.\vec{n}) dA \tag{1.64}$$

Since the flow is steady $\left(\frac{\partial}{\partial t} \iiint_{CV} \vec{V} \rho dv = 0\right)$, we get

$$T = \iint_{CS} \rho \vec{V}(\vec{V}.\vec{n}) dA \Rightarrow T = \dot{m}(V_{out} - V_{in}) \tag{1.65}$$

We write Eq. (1.65) for flow from area A_1 to area A_4 to get

$$T = \dot{m} V_f = \rho A_1 V_\infty V_f = \rho A_4 (V_\infty + V_f) V_f \tag{1.66}$$

We write the Bernoulli equation between areas A_1 and A_2 and between areas A_3 and A_4:

$$P_1 + \frac{1}{2}\rho V_\infty^2 = P_2 + \frac{1}{2}\rho(V_\infty + V_u)^2 \tag{1.67}$$

$$P_4 + \frac{1}{2}\rho(V_\infty + V_f)^2 = P_3 + \frac{1}{2}\rho(V_\infty + V_l)^2. \tag{1.68}$$

We do not use Bernoulli equation between 1 and 4 because energy and momentum are added by the rotor disk. Since $P_4 = P_1$ from Eqs. (1.67) and (1.68), we get

$$P_3 - P_2 = \frac{\rho}{2}\left(V_f^2 + 2V_f V_\infty\right) \tag{1.69}$$

We write the thrust (T) in terms of the pressure difference across the disk

$$T = (P_3 - P_2)A_2 \Rightarrow T = \frac{\rho}{2}A_2\left(V_f^2 + 2V_f V_\infty\right) \tag{1.70}$$

From Eqs. (1.66) and (1.70), we get

1.10 Momentum Theory for Axial Flight

$$\frac{\rho}{2}A_2\left(V_f^2 + 2V_f V_\infty\right) = \rho A_4 (V_\infty + V_f) V_f \tag{1.71}$$

From the equation of continuity, we write

$$\rho A_2 (V_\infty + V_u) = \rho A_4 (V_\infty + V_f) \tag{1.72}$$

From Eqs. (1.59) and (1.60), we get

$$V_f = 2V_u \tag{1.73}$$

or

$$w = 2v_i$$

where $v_i = V_u = V_l$ (induced velocity at the rotor plane) and $w = V_f$ (induced velocity at the slipstream). These velocities are induced by the rotor as it tries to create lift.

Induced velocity at the slipstream is two times the induced velocity at the rotor plane.

Substituting $V_\infty = 0$ in Eq. (1.70), we get the induced velocity for the hover case

$$v_h^2 = \frac{T}{2\rho A} \quad (A_2 = A_3 = A = \pi R^2 = \text{Rotor disk area}) \tag{1.74}$$

where R is the rotor radius.

1.11 Momentum Theory for Forward Flight

It is possible to extend the momentum theory concept to forward flight by making some assumption. Consider Fig. 1.17, which shows the forces acting on the rotor plane in forward flight.

Here, T is the thrust, P is the propulsive force, D is the drag, W is the weight of the helicopter, L is the lift, and w is the slipstream velocity.

We write the conservation of momentum equation in forward flight as

$$T = \iint_{CS} \rho \vec{V}(\vec{V}.\vec{n}) dA \Rightarrow T = \dot{m}(V_{\text{out}} - V_{\text{in}}) \tag{1.75}$$

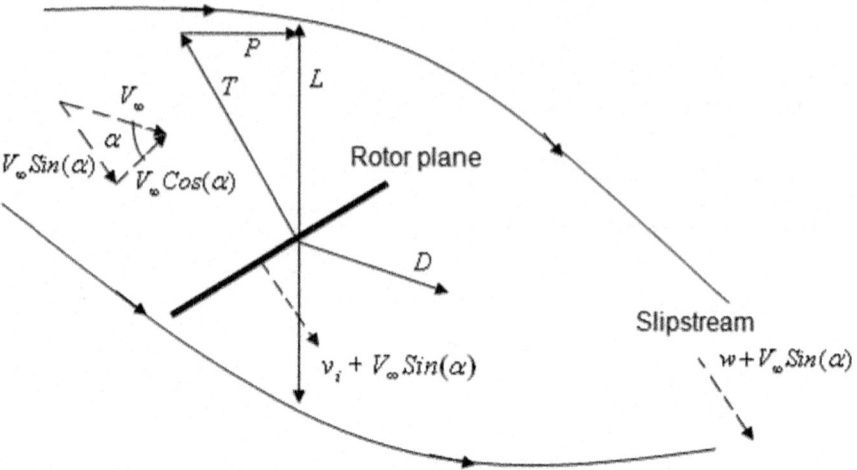

Fig. 1.17 Glauert flow model for momentum analysis of a rotor in forward flight

or

$$T = \dot{m}[w + V_\infty \sin(\alpha) - V_\infty \sin(\alpha)]$$

or

$$T = \dot{m}w \Rightarrow T = 2\dot{m}v_i \quad (1.76)$$

Here, $\dot{m} = \rho A U$ and $U = \sqrt{[V_\infty \cos(\alpha)]^2 + [v_i + V_\infty \sin(\alpha)]^2}$
We get

$$T = 2\rho A v_i \sqrt{[V_\infty \cos(\alpha)]^2 + [v_i + V_\infty \sin(\alpha)]^2}$$

or

$$v_i = \frac{T}{2\rho A \sqrt{[V_\infty \cos(\alpha)]^2 + [v_i + V_\infty \sin(\alpha)]^2}}$$

From Eq. (2.62), we get

$$v_i = \frac{v_h^2}{\sqrt{[V_\infty \cos(\alpha)]^2 + [v_i + V_\infty \sin(\alpha)]^2}} \quad (1.77)$$

Here, we introduce the terms inflow and advance ratio.

1.11 Momentum Theory for Forward Flight

Inflow (λ) is the non-dimensional coefficient of the velocity perpendicular to the plane of rotor:

$$\lambda_h = \frac{v_i}{\Omega R} \quad \text{(inflow in case of hover condition)} \qquad (1.78)$$

$$\lambda = \frac{V_\infty \sin(\alpha) + v_i}{\Omega R} \quad \text{(inflow in forward flight condition)} \qquad (1.79)$$

where V_∞ is the forward velocity, v_i is the induced velocity, Ω is the angular velocity, and α is the angle of attack.

Typically, non-dimensional form of equations is often used. Advance ratio (μ) is non-dimensional coefficient of the velocity parallel to the plane of the rotor:

$$\mu = \frac{V_\infty \cos(\alpha)}{\Omega R} \qquad (1.80)$$

From Eqs. (1.77), (1.79), and (1.80), we get

$$\lambda = \mu \tan(\alpha) + C_T/2\sqrt{(\mu^2 + \lambda^2)} \qquad (1.81)$$

where C_T is the non-dimensional coefficient of thrust:

$$C_T = \frac{T}{\rho A \Omega^2 R^2}$$

The model in (1.81) assumes that inflow is uniform. In reality, inflow is non-uniform and models have been developed to account for this reality. We can write an equation for inflow which will vary over the length of the blade and over the azimuthal angle:

$$\lambda = \frac{C_T/2}{\sqrt{\mu^2 + \lambda^2}} \left(1 + \frac{K_x x \cos(\psi)}{R} + \frac{K_y x \sin(\psi)}{R}\right) \quad \text{(linear inflow model)} \qquad (1.82)$$

where

$$K_x = \frac{4}{3}\left[(1 - 1.8\mu^2)\sqrt{1 + \left(\frac{\lambda}{\mu}\right)^2} - \frac{\lambda}{\mu}\right]$$

$$K_y = -2\mu.$$

The uniform inflow model is appropriate for hover and the linear inflow model for forward flight. In reality, the wake is highly non-uniform, i.e., $\lambda = \lambda(r, \psi)$, and free wake models are needed for its accurate prediction.

1.12 Newton–Raphson Method

The uniform inflow model for forward flight requires a numerical solution

$$f(\lambda) = \lambda - \mu \tan(\alpha) + C_T/2\sqrt{(\mu^2 + \lambda^2)} = 0$$

Assume the initial value of inflow to be λ_0

$$f(\lambda_0 + h) = 0$$

We write the Taylor series as

$$f(\lambda_0) + hf'(\lambda_0) + \frac{h^2}{2}f''(\lambda_0) + \text{Higher Order Terms} = 0$$

We assume the solution up to the first order

$$f(\lambda_0) + hf'(\lambda_0) = 0 \Rightarrow h = -\frac{f(\lambda_0)}{f'(\lambda_0)}$$

Thus,

$$\lambda_1 = \lambda_0 + h \Rightarrow \lambda_1 = \lambda_0 - \frac{f(\lambda_0)}{f'(\lambda_0)}$$

or

$$\lambda_{n+1} = \lambda_n - \frac{f(\lambda_n)}{f'(\lambda_n)} = \lambda_n - \frac{\lambda_n - \mu\tan(\alpha) + \frac{C_T}{2\sqrt{\mu^2 + \lambda_n^2}}}{1 - \frac{C_T \lambda_n}{2(\mu^2 + \lambda_n^2)^{3/2}}} \quad (1.83)$$

The hover value $\sqrt{\frac{C_T}{2}}$ can be used as the initial guess.

1.13 Blade Element Theory

Momentum theory considers the rotor to be an actuator disk and is not able to directly relate the blade section properties to the rotor thrust, power, etc. Blade element theory is useful for deriving equations which can guide blade design.

In this theory, thrust is calculated for a small section of the blade and then integrated over the length of the rotor blade to get the thrust produced by one blade.

1.13 Blade Element Theory

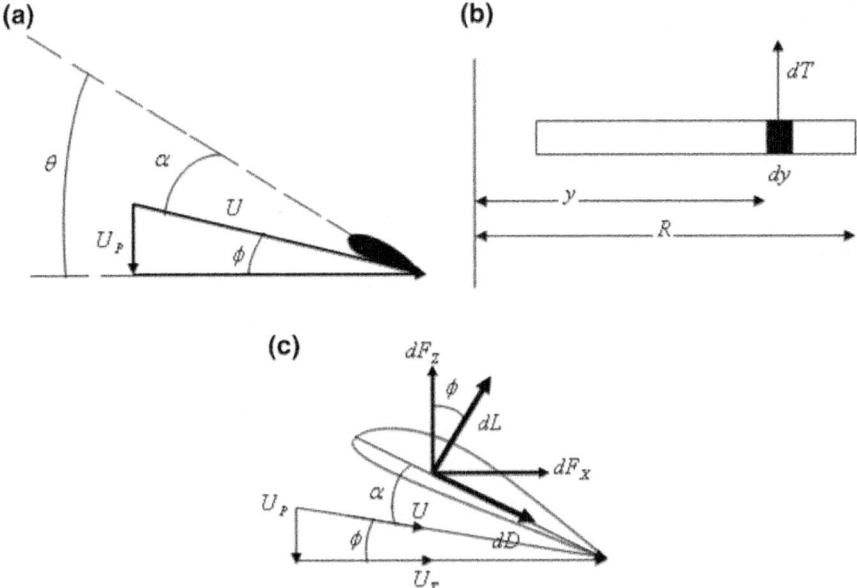

Fig. 1.18 a Blade element theory. b Blade element theory. c Blade element theory

Consider Fig. 1.18a, which shows the pitch angle (θ), inflow angle (φ), and effective angle of attack ($\alpha = \theta - \varphi$), where U_T and U_P are tangential and perpendicular components of flow velocity, respectively. Figure 1.18b shows an infinitesimal section dy; dT is the thrust produced by this section. Figure 1.18c shows the forces acting on the blade section. dL and dD are the lift and drag produced by the section and are perpendicular and parallel to the resultant velocity U, respectively. dF_Z and dF_X are the components of the force perpendicular and parallel to U_T.

We write the lift produced by the section dy

$$dL = \frac{1}{2}\rho U^2 c C_l dy \Rightarrow dL = \frac{1}{2}\rho (U_T^2 + U_P^2) c C_l dy \qquad (1.84)$$

We write the drag produced by the section dy

$$dD = \frac{1}{2}\rho U^2 c C_d dy \Rightarrow dD = \frac{1}{2}\rho (U_T^2 + U_P^2) c C_d dy \qquad (1.85)$$

We write the forces dF_Z and dF_X in terms of lift and drag

$$dF_Z = dL \cos(\phi) - dD \sin(\phi) \qquad (1.86)$$

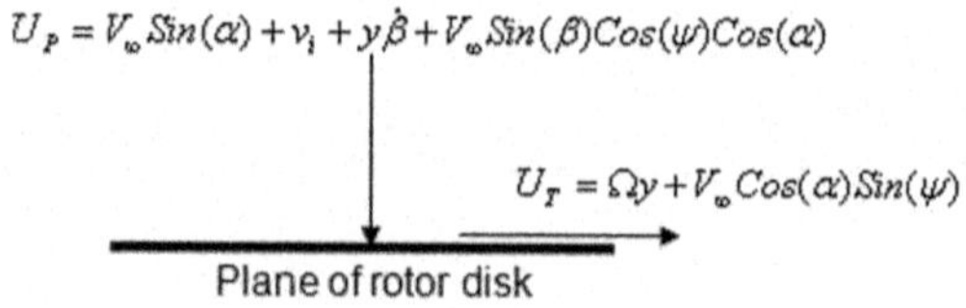

Fig. 1.19 Tangential and perpendicular components of the flow velocity

$$U_P = V_\infty Sin(\alpha) + v_i + y\dot{\beta} + V_\infty Sin(\beta)Cos(\psi)Cos(\alpha)$$

$$U_T = \Omega y + V_\infty Cos(\alpha) Sin(\psi)$$

Plane of rotor disk

or

$$dF_Z = \frac{1}{2}\rho(U_T^2 + U_P^2)cC_l dy \cos(\phi) - \frac{1}{2}\rho(U_T^2 + U_P^2)cC_d dy \sin(\phi) \quad (1.87)$$

$$dF_X = dL\sin(\phi) + dD\cos(\phi)$$

or

$$dF_X = \frac{1}{2}\rho(U_T^2 + U_P^2)cC_l dy \sin(\phi) + \frac{1}{2}\rho(U_T^2 + U_P^2)cC_d dy \cos(\phi) \quad (1.88)$$

Here, $dT = dF_Z$.

Consider Fig. 1.19, where tangential and perpendicular components of the flow velocity are shown.

$$U_T = \Omega y + V_\infty \cos(\alpha)\sin(\psi) \quad (1.89)$$

where Ωy is the result of angular velocity of rotor and $V_\infty \cos(\alpha)\sin(\psi)$ is the result of flow velocity:

$$U_P = V_\infty \sin(\alpha) + v_i + y\dot{\beta} + V_\infty \sin(\beta)\cos(\psi)\cos(\alpha) \quad (1.90)$$

where $V_\infty \sin(\alpha)$ is the result of flow velocity, v_i is the induced velocity, $y\dot{\beta}$ is the result of flap motion, and $V_\infty \sin(\beta)\cos(\psi)\cos(\alpha)$ is the result of anhedral effect.

We write Eqs. (1.89) and (1.90) as

$$U_T = \Omega R\left(\frac{y}{R} + \mu \sin(\psi)\right) \quad (1.91)$$

$$U_P = \Omega R\left(\lambda + \frac{y\dot{\beta}}{\Omega R} + \mu\beta\cos(\psi)\right). \quad (1.92)$$

These are the tangential and perpendicular velocities in forward flight. The expressions for velocities will be useful for deriving the helicopter rotor blade equations.

1.14 Derivation of Equation of Motion of Flapping Rigid Blade

Consider Fig. 1.20, which shows a rigid rotor blade, hinged at the root. The blade flaps up and down under the aerodynamic forces, centrifugal forces, and inertial forces. A small element having mass dm is taken at a distance y from the center. Here, β is the flapping angle, dL is the elemental lift force, Ω is the angular velocity, dF_Z is the force acting perpendicular to the rotor disk plane, and dC_F is the centrifugal force.

Each force is analyzed physically to form the equation of motion of the flapping rigid blade.

(a) Inertia Force

Force on the small segment having mass $dm = (y\ddot{\beta})dm$

Associated moment at the hinge $= y(y\ddot{\beta})dm$

Integrating all the small segments, moment due to full rotor $= I\ddot{\beta}$ \hfill (1.93)

where $I = \int_0^R dm y^2$, is the mass moment of inertia.

(b) Centrifugal Force

Force on the small segment having mass $dm = \frac{(\Omega y)^2 dm}{y} = \Omega^2 y dm$

Associated moment at the hinge $= \Omega^2 y dm (y \sin(\beta)) = \Omega^2 y^2 dm \beta$

where, $(y \sin(\beta) \approx y\beta)$

Fig. 1.20 Forces acting on a small element of a rigid rotor blade

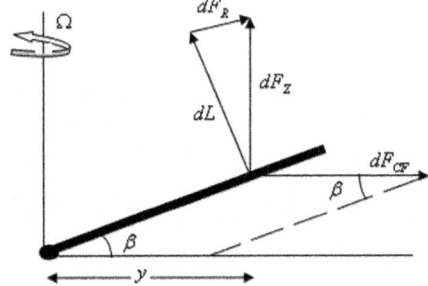

Integrating all the small segments, moment due to full rotor $= \int_0^R \Omega^2 y^2 \beta dm = I\Omega^2 \beta$

(1.94)

(c) Aerodynamic Force

Force on the small segment having mass $dm = Ldy$

Associated moment at the hinge $= Ldy(y\cos(\beta)) = Lydy$

where $(y\cos(\beta) \approx y)$

Integrating all the small segments, moment due to full rotor $= \int_0^R Lydy.$ (1.95)

We write moment equation about the hinge to get

$$I\ddot{\beta} + I\Omega^2\beta = \int_0^R Lydy \qquad (1.96)$$

Since $\psi = \Omega t$

$$\dot{\beta} = \frac{d\beta}{dt} = \frac{d\beta}{d\psi}\frac{d\psi}{dt} = \Omega\frac{d\beta}{d\psi}$$

$$\ddot{\beta} = \frac{d\dot{\beta}}{dt} = \frac{d\dot{\beta}}{d\psi}\frac{d\psi}{dt} = \Omega\frac{d\dot{\beta}}{d\psi} = \Omega\frac{d}{d\psi}\left(\Omega\frac{d\beta}{d\psi}\right) = \Omega^2\frac{d^2\beta}{d\psi^2}$$

We write Eq. (1.84) as

$$I\Omega^2\frac{d^2\beta}{d\psi^2} + I\Omega^2\beta = \int_0^R Lydy$$

or

$$\frac{d^2\beta}{d\psi^2} + \beta = \frac{1}{I\Omega^2}\int_0^R Lydy \qquad (1.97)$$

1.14 Derivation of Equation of Motion of Flapping Rigid Blade

where aerodynamic force per unit length is given by

$$L = \frac{1}{2}\rho U_T^2 c C_{L_\alpha}\left(\theta - \frac{U_P}{U_T}\right) \quad (1.98)$$

Here, θ is the pitch angle, c is the chord length, C_{L_α} is the lift curve slope, and U_T and U_P are tangential and perpendicular components of the flow velocity.

$$L = \frac{1}{2}\rho c C_{L_\alpha}\left(\theta U_T^2 - U_P U_T\right)$$

Substituting lift term in right-hand side of Eq. (1.97), we get

$$\frac{1}{I\Omega^2}\int_0^R Lydy = \frac{1}{I\Omega^2}\int_0^R \frac{1}{2}\rho c C_{L_\alpha}\left(\theta U_T^2 - U_P U_T\right)ydy \quad (1.99)$$

or

$$\frac{1}{I\Omega^2}\int_0^R Lydy = \frac{1}{I\Omega^2}\int_0^R \frac{1}{2}\rho c C_{L_\alpha}\theta U_T^2 ydy - \frac{1}{I\Omega^2}\int_0^R \frac{1}{2}\rho c C_{L_\alpha} U_P U_T ydy$$

or

$$\frac{1}{I\Omega^2}\int_0^R Lydy = \text{Term1} - \text{Term2} \quad (1.100)$$

Here, $\text{Term1} = \frac{1}{I\Omega^2}\int_0^R \frac{1}{2}\rho c C_{L_\alpha}\theta U_T^2 ydy$ and $\text{Term2} = \frac{1}{I\Omega^2}\int_0^R \frac{1}{2}\rho c C_{L_\alpha} U_P U_T ydy$

We write

$$\text{Term1} = \frac{1}{I\Omega^2}\int_0^R \frac{1}{2}\rho c C_{L_\alpha}\theta U_T^2 ydy \quad (1.101)$$

Here, $\theta = \theta_1 + \theta_{tw}\frac{y}{R}$, where $\theta_1 = \theta_0 + \theta_{1s}\sin(\psi) + \theta_{1c}\cos(\psi)$,

The helicopter is controlled through θ_0, θ_{1s}, and θ_{1c} which are inputs by the pilot to the main rotor via the swashplate. Here, θ_0 is called the collective pitch, θ_{1c} the lateral cyclic, and θ_{1s} the longitudinal cyclic. Also, θ_{tw} represents the built-in twist in the rotor blade.

From Eqs. (1.91) and (1.101), we get

$$\text{Term1} = \frac{1}{I\Omega^2} \frac{1}{2} \rho c C_{L_\alpha} \int_0^R \left\{ \Omega^2 R^2 \theta_1 \left[\frac{y^3}{R^2} + y\mu^2 \sin^2(\psi) + \frac{2y^2 \mu \sin(\psi)}{R} \right] \right.$$
$$\left. + \Omega^2 R^2 \theta_{tw} \left[\frac{y^4}{R^3} + \frac{y^2 \mu^2 \sin^2(\psi)}{R} + \frac{2y^3 \mu \sin(\psi)}{R^2} \right] \right\} dy$$

or

$$\text{Term1} = \frac{\rho c C_{L_\alpha} R^4}{I} \left\{ \theta_1 \left[\frac{1}{8} + \frac{\mu^2 \sin^2(\psi)}{4} + \frac{\mu \sin(\psi)}{3} \right] \right.$$
$$\left. + \theta_{tw} \left[\frac{1}{10} + \frac{\mu^2 \sin^2(\psi)}{6} + \frac{\mu \sin(\psi)}{4} \right] \right\}$$

Here, we define a term, Lock number $\gamma = \frac{\rho c C_{L_\alpha} R^4}{I}$, which represents the ratio of aerodynamic forcing to inertial forcing:

$$\text{Term1} = \gamma \left\{ \theta_1 \left[\frac{1}{8} + \frac{\mu^2 \sin^2(\psi)}{4} + \frac{\mu \sin(\psi)}{3} \right] \right.$$
$$\left. + \theta_{tw} \left[\frac{1}{10} + \frac{\mu^2 \sin^2(\psi)}{6} + \frac{\mu \sin(\psi)}{4} \right] \right\} \quad (1.102)$$

We write

$$\text{Term2} = \frac{1}{I\Omega^2} \int_0^R \frac{1}{2} \rho c C_{L_\alpha} U_P U_T y \, dy \quad (1.103)$$

From Eqs. (1.91), (1.92), and (1.103), we get

$$\text{Term2} = \frac{1}{I\Omega^2} \frac{1}{2} \rho c C_{L_\alpha} \int_0^R \Omega^2 R^2 \left\{ \lambda \left[\frac{y^2}{R} + y\mu \sin(\psi) \right] \right.$$
$$\left. + \dot{\beta} \left[\frac{y^3}{\Omega R^2} + \frac{y^2 \mu \sin(\psi)}{\Omega R} \right] + \beta \left[\frac{y^2 \mu \cos(\psi)}{R} + \mu^2 y \sin(\psi) \cos(\psi) \right] \right\} dy$$

or

$$\text{Term2} = \frac{\rho c C_{L_a} R^4}{I} \left\{ \lambda \left[\frac{1}{6} + \frac{\mu \sin(\psi)}{4} \right] + \dot{\beta} \left[\frac{1}{8\Omega} + \frac{\mu \sin(\psi)}{6\Omega} \right] \right. $$
$$\left. + \beta \mu \cos(\psi) \left[\frac{1}{6} + \frac{\mu \sin(\psi)}{4} \right] \right\}$$

or

$$\text{Term2} = \gamma \left\{ \lambda \left[\frac{1}{6} + \frac{\mu \sin(\psi)}{4} \right] + \frac{d\beta}{d\psi} \left[\frac{1}{8} + \frac{\mu \sin(\psi)}{6} \right] \right.$$
$$\left. + \beta \mu \cos(\psi) \left[\frac{1}{6} + \frac{\mu \sin(\psi)}{4} \right] \right\} \tag{1.104}$$

From Eqs. (1.102), (1.104), and (1.100), we get

$$\frac{1}{I\Omega^2} \int_0^R L y \, dy = \gamma \left\{ \theta_1 \left[\frac{1}{8} + \frac{\mu^2 \sin^2(\psi)}{4} + \frac{\mu \sin(\psi)}{3} \right] \right.$$
$$\left. + \theta_{\text{tw}} \left[\frac{1}{10} + \frac{\mu^2 \sin^2(\psi)}{6} + \frac{\mu \sin(\psi)}{4} \right] \right\}$$
$$- \gamma \left\{ \lambda \left[\frac{1}{6} + \frac{\mu \sin(\psi)}{4} \right] + \frac{d\beta}{d\psi} \left[\frac{1}{8} + \frac{\mu \sin(\psi)}{6} \right] \right.$$
$$\left. + \beta \mu \cos(\psi) \left[\frac{1}{6} + \frac{\mu \sin(\psi)}{4} \right] \right\}$$

or

$$\frac{1}{I\Omega^2} \int_0^R L y \, dy = \gamma \bar{M}_\beta \tag{1.105}$$

where

$$\bar{M}_\beta = \theta_1 \left\{ \frac{1}{8} + \frac{\mu}{3} \sin(\psi) + \frac{\mu^2}{4} \sin^2(\psi) \right\}$$
$$+ \theta_{\text{tw}} \left\{ \frac{1}{10} + \frac{\mu^2 \sin^2(\psi)}{6} + \frac{\mu \sin(\psi)}{4} \right\}$$
$$- \lambda \left\{ \frac{1}{6} + \frac{\mu}{4} \sin(\psi) \right\} - \frac{d\beta}{d\psi} \left\{ \frac{1}{8} + \frac{\mu}{6} \sin(\psi) \right\}$$
$$- \beta \mu \cos(\psi) \left\{ \frac{1}{6} + \frac{\mu}{4} \sin(\psi) \right\}$$

From Eqs. (1.105) and (1.97), we get

$$\frac{d^2\beta}{d\psi^2} + \beta = \gamma \bar{M}_\beta \tag{1.106}$$

If we put $\gamma = 0$, the above equation simulates a flapping blade in a vacuum. Solution to Eq. (1.106) is given by Fourier series

$$\beta(\psi) = \beta_0 + \sum_{n=1}^{N} (\beta_{nc} \cos(n\psi) + \beta_{ns} \sin(n\psi)) \tag{1.107}$$

Fourier series expansion for a given N is put into both sides of the equation, and the harmonic coefficients are equated. This approach is called the harmonic balance method.

1.15 Derivation of Elastic Rotor Blade Equation

Here, we derive the equation for free vibration of the rotor blade.

Consider Fig. 1.21a, which shows the deflection of an elastic blade. Figure 1.21b shows the forces acting on a small section of the elastic blade. Here, M is the bending moment, G is the centrifugal force, and S is the shear force acting on the blade section.

We write the force and moment balance equations, considering the centrifugal force and inertial force:

$$dG + m\Omega^2 x dx = 0 \Rightarrow G = \int_x^R m\Omega^2 x dx \tag{1.108}$$

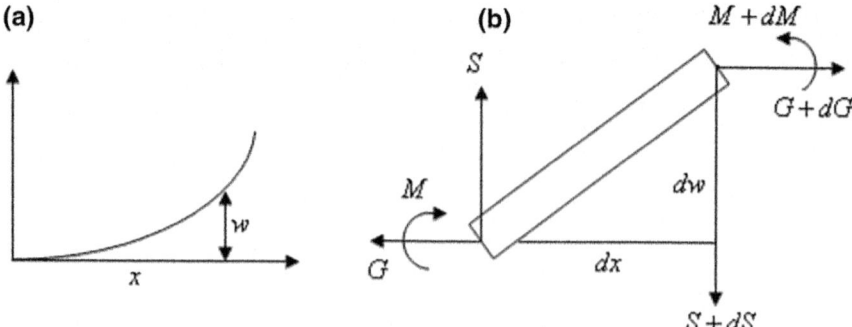

Fig. 1.21 a Deflection of an elastic rotor blade. b Force diagram

1.15 Derivation of Elastic Rotor Blade Equation

$$dS + mdx\frac{\partial^2 w}{\partial t^2} = 0 \Rightarrow \frac{\partial S}{\partial x} = -m\frac{\partial^2 w}{\partial t^2} \tag{1.109}$$

$$Gdw + Sdx - dM = 0 \Rightarrow \frac{\partial M}{\partial x} = S + G\frac{\partial w}{\partial x} \Rightarrow \frac{\partial^2 M}{\partial x^2} = \frac{\partial S}{\partial x} + \frac{\partial}{\partial x}\left(G\frac{\partial w}{\partial x}\right) \tag{1.110}$$

From Euler–Bernoulli beam theory, we can write

$$M = EI\frac{\partial^2 w}{\partial x^2} \tag{1.111}$$

From Eqs. (1.109) and (1.110), we write

$$\frac{\partial^2 M}{\partial x^2} = -m\frac{\partial^2 w}{\partial t^2} + \frac{\partial}{\partial x}\left(G\frac{\partial w}{\partial x}\right) \tag{1.112}$$

From Eqs. (1.111) and (1.112), we write

$$\frac{\partial^2}{\partial x^2}\left(EI\frac{\partial^2 w}{\partial x^2}\right) = -m\frac{\partial^2 w}{\partial t^2} + \frac{\partial}{\partial x}\left(G\frac{\partial w}{\partial x}\right) \tag{1.113}$$

From Eqs. (1.108) and (1.113), we write

$$\frac{\partial^2}{\partial x^2}\left(EI\frac{\partial^2 w}{\partial x^2}\right) = -m\frac{\partial^2 w}{\partial t^2} + \frac{\partial}{\partial x}\left(\left(\int_x^R m\Omega^2 x dx\right)\frac{\partial w}{\partial x}\right) \tag{1.114}$$

We write

$$\frac{\partial}{\partial x}\left(\left(\int_x^R m\Omega^2 x dx\right)\frac{\partial w}{\partial x}\right) = \Omega^2\left(-mx\frac{\partial w}{wx} + \frac{\partial^2 w}{\partial x^2}\int_x^R mx dx\right) \tag{1.115}$$

From Eqs. (1.114) and (1.115), we write the equation where we have shifted from t to ψ for the time coordinate:

$$(EIw'')'' + m\Omega^2\ddot{w} + \Omega^2\left[mxw' - w''\int_x^R mx dx\right] = 0 \tag{1.116}$$

Forced vibration equation is given by

$$(EIw'')'' + m\Omega^2 \ddot{w} + \Omega^2 \left[mxw' - w'' \int_x^R mx dx \right] = F(x, \psi) \tag{1.117}$$

where $(\dot{w} = \partial w/\partial \psi, w' = \partial w/\partial x)$

Forcing term is given by

$$F(x, \psi) = \frac{\rho a c}{2} * (U_T^2 \theta - U_P U_T) \tag{1.118}$$

Here, $U_T = \Omega R \left[\left(\frac{x}{R}\right) + \mu \sin(\psi) \right]$, $U_P = \Omega R \left[\left(\frac{\dot{w}}{R}\right) + \lambda + \mu w' \cos(\psi) \right]$, and $\theta = \theta_1 + \theta_{tw} \frac{x}{R}$, where $\theta_1 = \theta_0 + \theta_{1s} \sin(\psi) + \theta_{1c} \cos(\psi)$,

[Note that $\left(\dot{w} = \frac{\partial w}{\partial \psi}\right)$]

Substituting values of U_P, U_T, and θ in Eq. (1.118), we get

$$F(x, \psi) = \text{Const} \left\{ x^2 \frac{\theta_1}{R^2} + x \frac{2\theta_1 \mu \sin(\psi)}{R} \right.$$
$$- x \frac{\lambda}{R} + [\theta_1 \mu^2 \sin^2(\psi) - \lambda \mu \sin(\psi)] + x^3 \frac{\theta_{tw}}{R^3} + x \frac{\theta_{tw} \mu^2 \sin^2(\psi)}{R}$$
$$\left. + x^2 \frac{2\theta_{tw} \mu \sin(\psi)}{R^2} \right\} + \text{Const} \left\{ x \frac{-1}{R^2} + \frac{-\mu \sin(\psi)}{R} \right\} \frac{\partial w}{\partial \psi}$$
$$+ \text{Const} \left\{ x \frac{-\mu \cos(\psi)}{R} + \frac{-\mu^2 \sin(2\psi)}{2} \right\} \frac{\partial w}{\partial x} \tag{1.119}$$

We can see that the forcing term for the elastic blade equation has periodic terms such as $\sin(\psi), \sin^2(\psi)$ and $\sin(2\psi)$. These terms induce vibratory response and loads in the rotor system.

For the uniform inflow model, we can write

$$F(x, \psi) = a_{1x} a_{1\psi} + a_{2x} a_{2\psi} + a_{3x} a_{3\psi} + a_{4x} a_{4\psi} + a_{5x} a_{5\psi} + a_{6x} a_{6\psi}$$
$$+ a_{7x} a_{7\psi} + (b_{1x} b_{1\psi} + b_{2x} b_{2\psi}) \dot{w} + (c_{1x} c_{1\psi} + c_{2x} c_{2\psi}) w' \tag{1.120}$$

1.15 Derivation of Elastic Rotor Blade Equation

where Const $= \frac{\rho a c}{2}\Omega^2 R^2$,

$a_{1x} = \text{Const} * x^2, a_{2x} = \text{Const} * x, a_{3x} = \text{Const} * x, a_{4x} = \text{Const},$
$a_{5x} = \text{Const} * x^3, a_{6x} = \text{Const} * x, a_{7x} = \text{Const} * x^2,$
$a_{1\psi} = \theta_1/R^2, a_{2\psi} = 2\theta_1\mu\sin(\psi)/R, a_{3\psi} = -\lambda_i/R, a_{4\psi} = \theta_1\mu^2\sin^2(\psi) - \lambda_i\mu\sin(\psi),$
$a_{5\psi} = \theta_{tw}/R^3, a_{6\psi} = \mu^2\theta_{tw}\sin^2(\psi)/R, a_{7\psi} = 2\mu\theta_{tw}\sin(\psi)/R^2,$
$b_{1x} = \text{Const} * x, b_{2x} = \text{Const}, b_{1\psi} = -1/R^2, b_{2\psi} = -\mu\sin(\psi)/R,$
$c_{1x} = \text{Const} * x, c_{2x} = \text{Const}, c_{1\psi} = -\mu\cos(\psi)/R, \text{ and } c_{2\psi} = -\mu^2\sin(2\psi)/2$

Equation (1.120) shows the presence of motion-dependent forces. For example, forces are a function of $\dot{w}$ and w'. Such terms need to be taken to the left-hand side of Eq. (1.117), where they will influence the damping and stiffness of the rotor system. The presence of motion-dependent forces makes the rotor dynamic problem an aeroelastic problem.

Chapter 2
Finite Element Analysis in Space

2.1 Introduction

In this chapter, finite element in space is discussed in detail, and it is typically the first step in the solution of the elastic rotor problem as it yields the rotating natural frequencies. Bar, beam, and rotating beam finite element formulation are explained.

2.2 Finite Element in Space

Finite element method is a numerical method for getting approximate solution of differential equations. Analytical solution of the equation gives us the exact solution at any point in that domain, while finite element method gives us the approximate solution at discrete number of points in that domain. In finite element method, we divide the full domain into a number of elements, which are connected through nodal points. Then, we write equations for each element and combine them to get the solution. Finite element approach is a weak formulation of the physical problem.

2.3 Strong Form of the Equation

Consider Fig. 2.1, a case of an elastic bar subjected to uniform load.

The governing differential equation (2.1) along with boundary conditions (2.2) and (2.3) gives the strong form of the problem

$$EA\frac{d^2u}{dx^2} = f_0 \qquad (2.1)$$

Fig. 2.1 Elastic bar subjected to uniform load

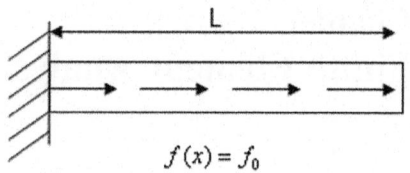

$$u(0) = 0 \quad (2.2)$$

$$EA\frac{du}{dx}\bigg|_{(x=L)} = 0 \quad (2.3)$$

where u is the axial displacement; Young's modulus (E) and cross sectional area (A) are constant over the length of the bar.

2.4 Weak Form of the Equation

The weak form is a variational statement of the equation, where we multiply the differential equation by a test function (v) and integrate it over the domain.

$$\int_0^L v\left(EA\frac{d^2u}{dx^2} - f_0\right)dx = 0 \quad (2.4)$$

We choose test function (v) such that it satisfies the homogeneous boundary conditions. Equation (2.4) can be written after integration by parts as

$$-\int_0^L EA\frac{du}{dx}\frac{dv}{dx} + \left[EAv\frac{du}{dx}\right]_0^L = \int_0^L f_0 v\, dx$$

or

$$-\int_0^L EA\frac{du}{dx}\frac{dv}{dx} + EAv(L)\frac{du}{dx}\bigg|_{x=L} - EAv(0)\frac{du}{dx}\bigg|_{x=0} = \int_0^L f_0 v\, dx$$

We apply boundary condition (2.2), (2.3), and $v(0) = 0$, to get the weak form of the equation

2.4 Weak Form of the Equation

$$-\int_0^L EA \frac{du}{dx}\frac{dv}{dx} = \int_0^L f_0 v\, dx \qquad (2.5)$$

The order of the derivatives in the equation is reduced in the weak form.

2.5 Galerkin's Method

In Galerkin's method, we find the solution $u = \tilde{u}$, such that

$$\int_0^L v\left(EA\frac{d^2\tilde{u}}{dx^2} - f_0\right)dx = 0 \qquad (2.6)$$

$$\tilde{u}(0) = 0 \qquad (2.7)$$

$$EA\frac{d^2\tilde{u}}{dx^2} = f_0 \qquad (2.8)$$

where we choose $\tilde{u}(x) = \sum_{j=1}^{N} c_j \phi_j(x)$ and $v(x) = \sum_{i=1}^{N} b_j \phi_j(x)$. Here, c_j is unknown and b_j is arbitrarily chosen. The interpolation or basis function $\phi_j(x)$ must satisfy all the boundary conditions for the problem. A good solution is obtained by taking many terms of the series. While Galerkin's method uses global interpolation function, the key idea in finite element method is to interpolate locally.

2.6 Shape Function in 1 Dimension

In finite element, we get the solution at nodal points of an element. Interpolation within the element is achieved by shape function.

Here, we take a bar element for illustration. Governing equation of linear-elastic bar element is

$$\frac{d}{dx}\left(AE\frac{du}{dx}\right) = 0 \qquad (2.9)$$

Consider Fig. 2.2, two-node bar element. Here q_1 and q_2 are the displacements at the two nodes and are called degrees of freedom.

Fig. 2.2 Bar element for shape function formulation

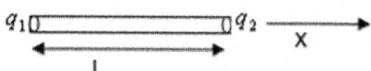

Linear displacement along the x-axis is assumed as

$$u(x) = a_0 + a_1 x \qquad (2.10)$$

At $x = 0$

$$u(0) = a_0 \Rightarrow q_1 = a_0 \qquad (2.11)$$

At $x = L$

$$u(L) = a_0 + a_1 L \Rightarrow q_2 = a_0 + a_1 L \qquad (2.12)$$

From (2.11) and (2.12), we get $a_0 = q_1, a_1 = \frac{q_2 - q_1}{L}$

We write the linear displacement within the finite element as $u(x) = q_1 + \left(\frac{q_2 - q_1}{L}\right)x \Rightarrow u(x) = q_1\left(1 - \frac{x}{L}\right) + q_2 \frac{x}{L} \Rightarrow u(x) = H_1 q_1 + H_2 q_2$
or

$$u(x) = [H_1 \; H_2] \begin{bmatrix} q_1 \\ q_2 \end{bmatrix} \qquad (2.13)$$

where $H_1 = \left(1 - \frac{x}{L}\right)$ and $H_2 = \frac{x}{L}$ are shape functions for the bar elements. Typically, polynomials are used as shape functions in finite element methods.

2.7 Shape Function Formulation for Beam Element

The static governing differential equation of an Euler–Bernoulli beam is given by

$$\frac{d^2}{dx^2}\left(EI \frac{d^2 v}{dx^2}\right) = 0 \qquad (2.14)$$

Here, in Fig. 2.3, we consider a beam element
Each element has two nodes; each node has vertical displacement and rotation.

Fig. 2.3 Beam element for shape function formulation

2.7 Shape Function Formulation for Beam Element

Total degree of freedom per element is 4. Assume the transverse displacement $v(x)$ to be

$$v(x) = a_0 + a_1 x + a_2 x^2 + a_3 x^3 \tag{2.15}$$

At $x = 0$

$$v(0) = a_0 \Rightarrow q_1 = a_0 \tag{2.16}$$

$$\frac{dv(0)}{dx} = a_1 \Rightarrow q_2 = a_1 \tag{2.17}$$

At $x = L$

$$v(L) = a_0 + a_1 L + a_2 L^2 + a_3 L^3 \Rightarrow q_3 = a_0 + a_1 L + a_2 L^2 + a_3 L^3 \tag{2.18}$$

$$\frac{dv(L)}{dx} = a_1 + 2a_2 L + 3a_3 L^2 \Rightarrow q_4 = a_1 + 2a_2 L + 3a_3 L^2 \tag{2.19}$$

From Eqs. (2.16)–(2.19), we write

$$\begin{bmatrix} q_1 \\ q_2 \\ q_3 \\ q_4 \end{bmatrix} = \begin{bmatrix} 1 & 0 & 0 & 0 \\ 0 & 1 & 0 & 0 \\ 1 & L & L^2 & L^3 \\ 0 & 1 & 2L & 3L^2 \end{bmatrix} \begin{bmatrix} a_0 \\ a_1 \\ a_2 \\ a_3 \end{bmatrix} \tag{2.20}$$

or

$$\begin{bmatrix} a_0 \\ a_1 \\ a_2 \\ a_3 \end{bmatrix} = \begin{bmatrix} 1 & 0 & 0 & 0 \\ 0 & 1 & 0 & 0 \\ \frac{-3}{L^2} & \frac{-2}{L} & \frac{3}{L^2} & \frac{-1}{L} \\ \frac{2}{L^3} & \frac{1}{L^2} & \frac{-2}{L^3} & \frac{1}{L^2} \end{bmatrix} \begin{bmatrix} q_1 \\ q_1 \\ q_3 \\ q_4 \end{bmatrix} \tag{2.21}$$

From Eq. (2.15), we write

$$v(x) = \begin{bmatrix} 1 & x & x^2 & x^3 \end{bmatrix} \begin{bmatrix} a_0 \\ a_1 \\ a_2 \\ a_3 \end{bmatrix} \tag{2.22}$$

or

$$v(x) = \begin{bmatrix} 1 & x & x^2 & x^3 \end{bmatrix} \begin{bmatrix} 1 & 0 & 0 & 0 \\ 0 & 1 & 0 & 0 \\ \frac{-3}{L^2} & \frac{-2}{L} & \frac{3}{L^2} & \frac{-1}{L} \\ \frac{2}{L^3} & \frac{1}{L^2} & \frac{-2}{L^3} & \frac{1}{L^2} \end{bmatrix} \begin{bmatrix} q_1 \\ q_1 \\ q_3 \\ q_4 \end{bmatrix} \tag{2.23}$$

or

$$v(x) = \left[2\left(\tfrac{x}{L}\right)^3 - 3\left(\tfrac{x}{L}\right)^2 + 1 \quad \tfrac{x^3}{L^2} - 2\tfrac{x^2}{L} + x \quad 3\left(\tfrac{x}{L}\right)^2 - 2\left(\tfrac{x}{L}\right)^3 \quad \tfrac{x^3}{L^2} - \tfrac{x^2}{L} \right] \begin{bmatrix} q_1 \\ q_1 \\ q_3 \\ q_4 \end{bmatrix} \quad (2.24)$$

or

$$v(x) = [H_1 \ H_2 \ H_3 \ H_4] \begin{bmatrix} q_1 \\ q_2 \\ q_3 \\ q_4 \end{bmatrix} = [H][q] \quad (2.25)$$

where H_1, H_2, H_3 and H_4 are shape functions for the beam finite element.

$$H_1 = 2\left(\frac{x}{L}\right)^3 - 3\left(\frac{x}{L}\right)^2 + 1, H_2 = \frac{x^3}{L^2} - 2\frac{x^2}{L} + x, H_3 = 3\left(\frac{x}{L}\right)^2 - 2\left(\frac{x}{L}\right)^3, H_4$$
$$= \frac{x^3}{L^2} - \frac{x^2}{L}.$$

2.8 Properties of Shape Function in 1D

In the rotor blade problem, we focus on 1D structures. The properties of the shape functions are discussed next.

1. Kronecker delta property

   Shape function of a node has value equal to one on that node and zero at all the other nodes.
   Consider Fig. 2.2.
   Node 1 $(x = 0)$

$$H_1 = 1 - \frac{x}{L} \Rightarrow H_1 = 1 \quad \text{and} \quad H_2 = \frac{x}{L} \Rightarrow H_2 = 0$$

Node 2 $(x = L)$

$$H_2 = \frac{x}{L} \Rightarrow H_2 = 1 \quad \text{and} \quad H_1 = 1 - \frac{x}{L} \Rightarrow H_1 = 0.$$

2. Compatibility condition

   Displacement approximation is continuous across element boundaries. Consider Fig. 2.4, where two elements are taken in a bar.

2.8 Properties of Shape Function in 1D

Fig. 2.4 Two elements in a bar for FEM in space

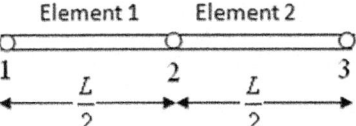

For formulation of shape function, we follow Eqs. (2.10)–(2.13). We substitute $x = 0$ and $x = L/2$ and get the shape function for first element $\begin{bmatrix} H_1^{(1)} & H_2^{(1)} \end{bmatrix}$, and we substitute $x = L/2$ and $x = L$ and get the shape function for second element $\begin{bmatrix} H_1^{(2)} & H_2^{(2)} \end{bmatrix}$.

$$u(x) = H_1^{(1)} q_1 + H_2^{(1)} q_2 = \left(1 - \frac{2x}{L}\right) q_1 + \frac{2x}{L} q_2 \qquad (2.26)$$

For second element, we write

$$u(x) = H_1^{(2)} q_2 + H_2^{(2)} q_3 = \left(2 - \frac{2x}{L}\right) q_2 + \left(\frac{2x}{L} - 1\right) q_3 \qquad (2.27)$$

Put $x = \frac{L}{2}$ into Eqs. (2.26) and (2.27) to get

$$u(x) = q_2.$$

3. Completeness

(a) Rigid body mode

$$H_1 + H_2 = 1$$

If the element moves by an unit displacement ($q_1, q_2 = 1$), displacement at any point in the element should be one.

$$u(x) = H_1 q_1 + H_2 q_2 = 1 \quad (\text{for } q_1, q_2 = 1)$$

(b) Constant strain state

Consider Fig. 2.5, if $q_1 = L/2$ and $q_2 = L$, then

$$\varepsilon(x) = \frac{q_1 - q_2}{\frac{L}{2}} = \frac{L - \frac{L}{2}}{\frac{L}{2}} = 1$$

Fig. 2.5 Bar element (shape function properties)

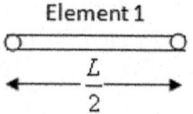

Check of strain state with displacement approximation

$$u(x) = H_1 q_1 + H_2 q_2 = \left(1 - \frac{2x}{L}\right) q_1 + \frac{2x}{L} q_2$$

$$u(x) = \left(1 - \frac{2x}{L}\right)\frac{L}{2} + \frac{2x}{L} L$$

or

$$u(x) = x + \frac{L}{2}$$

or

$$\varepsilon(x) = 1$$

A brief outline of finite element has been provided. We are now ready to apply the finite element method for the rotating beam problem.

2.9 Finite Element Formulation of Rotating Beam

Finite element formulation in space for the rotating beam is done using Hamilton's energy principle. (Complete derivation is given in [14].)

Potential energy is given by

$$V = \frac{1}{2}\int_0^R EI\left(\frac{\partial^2 w}{\partial x^2}\right)^2 dx + \frac{1}{2}\int_0^R G\left(\frac{\partial w}{\partial x}\right)^2 dx \qquad (2.28)$$

Kinetic energy is given by

$$T = \frac{1}{2}\int_0^R m\left(\frac{\partial w}{\partial t}\right)^2 dx \qquad (2.29)$$

where w is the transverse displacement, G is the centrifugal force, and m is the mass per unit length.

2.9 Finite Element Formulation of Rotating Beam

From Eq. (2.25), we have shape function of the beam element

$$w(x) = \begin{bmatrix} H_1 & H_2 & H_3 & H_4 \end{bmatrix} \begin{bmatrix} q_1 \\ q_2 \\ q_3 \\ q_4 \end{bmatrix} = [H][q]$$

or

$$\frac{\partial w}{\partial x} = \begin{bmatrix} H'_1 & H'_2 & H'_3 & H'_4 \end{bmatrix} \begin{bmatrix} q_1 \\ q_2 \\ q_3 \\ q_4 \end{bmatrix} = [H'][q]$$

or

$$\left(\frac{\partial^2 w}{\partial x^2}\right) = \begin{bmatrix} H''_1 & H''_2 & H''_3 & H''_4 \end{bmatrix} \begin{bmatrix} q_1 \\ q_2 \\ q_3 \\ q_4 \end{bmatrix} = [H''][q]$$

$$\left(\frac{\partial w}{\partial x}\right)^2 = [q]^T [H']^T [H'][q]$$

We write Eq. (2.28) as

$$V = \frac{1}{2}\int_0^R EI[q]^T[H'']^T[H''][q]\mathrm{d}x + \frac{1}{2}\int_0^R G[q]^T[H']^T[H'][q]\mathrm{d}x \tag{2.30}$$

or

$$V = \frac{1}{2}[q]^T \left(\int_0^R EI[H'']^T[H'']\mathrm{d}x + \int_0^R G[H']^T[H']\mathrm{d}x \right) [q] \tag{2.31}$$

or

$$V = \frac{1}{2}[q]^T_{1*4}[K]_{4*4}[q]_{4*1} \tag{2.32}$$

We write Eq. (2.29) as

$$T = \frac{1}{2}\int_0^R m[\dot{q}]^T[H]^T[H][q]\,dx \qquad (2.33)$$

or

$$T = [\dot{q}]^T \left(\frac{1}{2}\int_0^R m[H]^T[H]\,dx\right)[\dot{q}] \qquad (2.34)$$

or

$$T = [\dot{q}]^T_{1*4}[M]_{4*4}[\dot{q}]_{4*1} \qquad (2.35)$$

Rewriting the Lagrange's equation

$$\frac{d}{dt}\left(\frac{\partial L}{\partial \dot{q}_i}\right) - \frac{\partial L}{\partial q_i} = 0 \qquad (1.47)$$

where $L = T - V$. From Eqs. (2.32), (2.35), and (1.35), we get the free vibration problem

$$[M][\ddot{q}] + [K][q] = 0 \qquad (2.36)$$

where

$$[M] = \int_0^R m[H]^T[H]\,dx \qquad (2.37)$$

$$[K] = \left(\int_0^R EI[H'']^T[H'']\,dx + \int_0^R G[H']^T[H']\,dx\right) \qquad (2.38)$$

or

$$[K] = [K_1] + [K_2] \qquad (2.39)$$

2.9 Finite Element Formulation of Rotating Beam

Assuming EI and m constant over the length of the blade:
where

$$[K_1] = EI \begin{bmatrix} \int_0^R (H_1'')^2 dx & \int_0^R H_1'' H_2'' dx & \int_0^R H_1'' H_3'' dx & \int_0^R H_1'' H_4'' dx \\ & \int_0^R (H_2'')^2 dx & \int_0^R H_2'' H_3'' dx & \int_0^R H_2'' H_4'' dx \\ & & \int_0^R (H_3'')^2 dx & \int_0^R H_3'' H_4'' dx \\ & & & \int_0^R (H_4'')^2 dx \end{bmatrix},$$

$$[K_2] = \begin{bmatrix} \int_0^R G(H_1')^2 dx & \int_0^R GH_1' H_2' dx & \int_0^R GH_1' H_3' dx & \int_0^R GH_1' H_4' dx \\ & \int_0^R G(H_2')^2 dx & \int_0^R GH_2' H_3' dx & \int_0^R GH_2' H_4' dx \\ & & \int_0^R G(H_3')^2 dx & \int_0^R GH_3' H_4' dx \\ & & & \int_0^R G(H_4')^2 dx \end{bmatrix},$$

$$[M] = m \begin{bmatrix} \int_0^R (H_1)^2 dx & \int_0^R H_1 H_2 dx & \int_0^R H_1 H_3 dx & \int_0^R H_1 H_4 dx \\ & \int_0^R (H_2)^2 dx & \int_0^R H_2 H_3 dx & \int_0^R H_2 H_4 dx \\ & & \int_0^R (H_3)^2 dx & \int_0^R H_3 H_4 dx \\ & & & \int_0^R (H_4)^2 dx \end{bmatrix}$$

We solve the free vibration problem with the above matrices. The formulation is valid for any general axial force $G(x)$. For a rotating beam, we are interested in the centrifugal force.

2.10 Centrifugal Force

Centrifugal force is the additional term to beam equation in rotating beam equation. In the formulation, it is $[K_2]$ (Fig. 2.6).

Fig. 2.6 Centrifugal force on the rotating beam

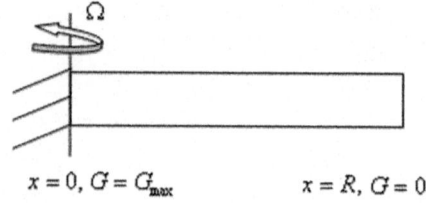

$$G = \int_x^R m\Omega^2 x \, dx \tag{2.40}$$

For constant mass per unit length

$$G = m\Omega^2 \left(\frac{R^2}{2} - \frac{x^2}{2} \right). \tag{2.41}$$

2.11 Shape Function Formulation for Two Elements

Consider Fig. 2.7, shape function for both the elements will be different.

For formulation of shape function, we follow Eqs. (2.15)–(2.25). For first element, we evaluate value of x at 0 and $L/2$. For second element, we evaluate value of x at $L/2$ and L.

Shape function for element 1 ($[H]_1$)	Shape function for element 2 ($[H]_2$)
$16\frac{x^3}{L^3} - 12\frac{x^2}{L^2} + 1,$	$24\frac{x}{L} - 36\frac{x^2}{L^2} + 16\frac{x^3}{L^3} - 4,$
$x - 4\frac{x^2}{L} + 4\frac{x^3}{L^2},$	$8x - 2L - 10\frac{x^2}{L} + 4\frac{x^3}{L^2},$
$12\frac{x^2}{L^2} - 16\frac{x^3}{L^3},$	$36\frac{x^2}{L^2} - 24\frac{x}{L} - 16\frac{x^3}{L^3} + 5,$
$4\frac{x^3}{L^2} - 2\frac{x^2}{L}$	$5x - L - 8\frac{x^2}{L} + 4\frac{x^3}{L^2}$

We integrate shape functions over the domain

Element 1 $\left(\int_0^{L/2} [H]_1^T dx \right)$	Element 2 $\left(\int_{L/2}^{L} [H]_2^T dx \right)$
$\frac{L}{4},$	$\frac{L}{4}$
$\frac{L^2}{48},$	$\frac{L^2}{48}$
$\frac{L}{4}$	$\frac{L}{4}$
$-\frac{L^2}{48}$	$-\frac{L^2}{48}$

2.11 Shape Function Formulation for Two Elements

Fig. 2.7 Shape function formulation of two elements

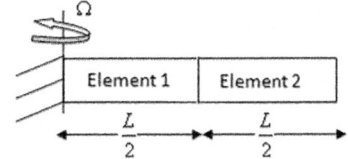

We see that

$$\int_0^{L/2} [H]_1^T dx = \int_{L/2}^{L} [H]_2^T dx \qquad (2.42)$$

We write $[K_1]$ matrix

$$[K_1] = EI \int_0^R [H'']^T [H''] dx$$

We notice that

$$[K_1]\text{1st(element)} = [K_1]\text{2nd(element)}$$

or

$$EI \int_0^{L/2} [H'']_1^T [H'']_1 dx = EI \int_{L/2}^{L} [H'']_2^T [H'']_2 dx \qquad (2.43)$$

So, we should calculate the stiffness matrix $[K_1]$ for only one element, and it will be the same for all the elements.

Now we integrate $\int_0^{L/2} [H]_1^T f(x) dx$ and $\int_{L/2}^{L} [H]_2^T f(x) dx$ where $f(x) = x$

Element 1 $\left(\int_0^{L/2} [H]_1^T x dx\right)$	Element 2 $\left(\int_{L/2}^{L} [H]_2^T x dx\right)$
$\frac{3L^2}{80}$	$\frac{13L^2}{80}$
$\frac{L^3}{240}$	$\frac{7L^3}{480}$
$\frac{7L^2}{80}$	$\frac{17L^2}{80}$
$-\frac{L^3}{160}$	$-\frac{L^3}{60}$

We see that

$$\int_0^{L/2} [H]_1^T x\,dx \neq \int_{L/2}^L [H]_2^T x\,dx \tag{2.44}$$

We write $[K_2]$ matrix

$$[K_2] = \int_0^R G[H']^T [H']\,dx$$

We notice that

$$[K_2]1\text{st}(\text{element}) \neq [K_2]2\text{nd}(\text{element})$$

or

$$\int_0^{L/2} G(x)[H']_1^T [H']_1 \, dx \neq \int_{L/2}^L G(x)[H']_2^T [H']_2 \, dx \tag{2.45}$$

So, we should calculate the stiffness matrix $[K_2]$ for each element.

2.12 FEM Formulation of Rotating Beam with Only One Shape Function (for Free Vibration)

Formulation of $[K_2]$ matrix with the shape function of the first element

Considering Fig. 2.8, we write the FEM formulation for third element using the shape function of one element. Here, L is the length of each element, R is the radius of rotor, x is the length along the element, x_i is the distance of element from the starting point and will depend on the element we have taken, Ω is the angular velocity, N is the number of elements, ρ is the global coordinate system, and x is the local coordinate system.

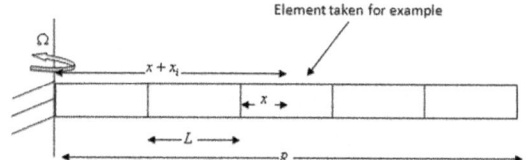

Fig. 2.8 FEM formulation of rotating beam with the shape function of one element (for free vibration)

2.12 FEM Formulation of Rotating Beam with Only One ...

EI is stiffness, and m_i is mass of particular element.
Rewriting $[K_2]$ matrix

$$[K_2] = \int_0^l G(x)[H']_1^T [H']_1 dx$$

From Fig. 3.9, we write the centrifugal force for the element as

$$G(x) = \int_{x_i+x}^{R} m\Omega^2 \rho d\rho \Rightarrow G(x) = \int_{x_i}^{R} m\Omega^2 \rho d\rho - \int_{x_i}^{x+x_i} m\Omega^2 \rho d\rho$$

or

$$G(x) = \text{Term1} - \text{Term2} \quad (2.46)$$

$$\text{Term1} = \int_{x_i}^{R} m\Omega^2 \rho d\rho \Rightarrow \text{Term1} = \sum_{j=i}^{N} \int_{x_j}^{x_{j+1}} m\Omega^2 x dx$$

or

$$\text{Term1} = \sum_{j=i}^{N} m_j \Omega^2 \frac{(x_{j+1}^2 - x_j^2)}{2} \Rightarrow \text{Term1} = \Omega^2 \frac{A_i}{2}$$

where $A_i = \sum_{j=i}^{N} m_j \frac{(x_{j+1}^2 - x_j^2)}{2}$ (It will be a constant term)
We write Term2

$$\text{Term2} = \int_{x_i}^{x+x_i} m\Omega^2 \rho d\rho \Rightarrow \text{Term2} = \frac{m_i \Omega^2}{2} \{(x_i+x)^2 - x_i^2\}$$

or

$\text{Term2} = \frac{m_i \Omega^2}{2}(2xx_i + x^2)$ (It will be a varying term because of x)
We write Eq. (2.46) as

$$G(x) = \frac{\Omega^2 A_i}{2} - \frac{m_i \Omega^2}{2}(2xx_i + x^2) \quad (2.47)$$

Now we rewrite $[K_2]$

$$[K_2] = \int_0^l G(x)[H']_1^T[H']_1 \, dx$$

$$[K_2]_i = \int_0^l \left\{ \frac{\Omega^2 A_i}{2} - \frac{m_i \Omega^2}{2}(2xx_i + x^2) \right\} [H']_1^T [H']_1 \, dx \qquad (2.48)$$

where $i = 1, 2, 3$ (for different elements).

From this formulation, we get

$$[M] = m \begin{bmatrix} \frac{13l}{35} & \frac{11l^2}{210} & \frac{9l}{70} & \frac{-13l^2}{420} \\ & \frac{l^3}{105} & \frac{13l^2}{420} & \frac{-l^3}{140} \\ & & \frac{13l}{35} & \frac{-11l^2}{210} \\ & & & \frac{l^3}{105} \end{bmatrix},$$

$$[K_1] = EI \begin{bmatrix} \frac{12}{l^3} & \frac{6}{l^2} & \frac{-12}{l^3} & \frac{6}{l^2} \\ & \frac{4}{l} & \frac{-6}{l^2} & \frac{2}{l} \\ & & \frac{12}{l^3} & \frac{-6}{l^2} \\ & & & \frac{4}{l} \end{bmatrix}$$

$$[K_2]_i = \frac{\Omega^2 A_i}{2} \begin{bmatrix} \frac{6}{5l} & \frac{1}{10} & \frac{-6}{5l} & \frac{1}{10} \\ & \frac{2l}{25} & \frac{-1}{10} & \frac{-l}{30} \\ & & \frac{6}{5l} & \frac{-1}{10} \\ & & & \frac{2l}{15} \end{bmatrix}$$

$$- m_i \Omega^2 \begin{bmatrix} \frac{3x_i}{5} + \frac{6l}{35} & \frac{x_il}{10} + \frac{l^2}{28} & \frac{-3x_i}{5} - \frac{6l}{35} & -\frac{l^2}{70} \\ & \frac{l^2 x_i}{30} + \frac{l^3}{105} & \frac{-lx_i}{10} - \frac{l^2}{28} & \frac{-l^2 x_i}{60} - \frac{l^3}{70} \\ & & \frac{3x_i}{5} + \frac{6l}{35} & \frac{l^2}{70} \\ & & & \frac{l^2 x_i}{10} + \frac{3l^2}{70} \end{bmatrix}$$

When $\Omega = 0$, $[K_2] = [0]$ and the formulation reduces to the well-known non-rotating beam element found in [15]. When $\Omega \neq 0$, the centrifugal stiffening causes a spatially varying stiffness matrix $[K_2]$.

2.13 Calculation of Mode Shapes and Frequencies

We rewrite Eq. (2.36)

$$[M][\ddot{q}] + [K][q] = 0 \text{ (Free vibration problem)}$$

The natural frequency f and the respective mode shape V of a rotating beam can be obtained from the Jacobian matrix $[A]$.

$$[A] = [M]^{-1}[K] \tag{2.49}$$

$$f = \sqrt{\text{eigval}([A])} \tag{2.50}$$

$$V = \text{eigvec}([A]) \tag{2.51}$$

Non-dimensional rotating frequency η and non-dimensional rotating speed s are given by

$$s = \Omega\sqrt{\frac{mR^4}{EI}}, \eta = f\sqrt{\frac{mR^4}{EI}}$$

We have solved the free vibration problem, so we consider the case of forced vibration.

$$[M][\ddot{q}] + [K][q] = \int_0^L [H]^T F(x, \psi) dx \tag{2.52}$$

where $\dot{q} = dq/dt$.

2.14 FEM Formulation of Aerodynamic Force for Rotor Problem

We have done finite element formulation for the free vibration. We develop the finite element formulation for the forced vibration in this section.

Finite element formulation of the right-hand side of Eq. (1.105) yields

$$[Q] = \int_0^L [H]^T F(x, \psi) dx \text{ (Element load vector)} \tag{2.53}$$

We rewrite Eq. (1.108)

$$F(x, \psi) = a_{1x}a_{1\psi}$$
$$+ a_{2x}a_{2\psi} + a_{3x}a_{3\psi} + a_{4x}a_{4\psi} + a_{5x}a_{5\psi} + a_{6x}a_{6\psi} + a_{7x}a_{7\psi}$$
$$+ (b_{1x}b_{1\psi} + b_{2x}b_{2\psi})\dot{w} + (c_{1x}c_{1\psi} + c_{2x}c_{2\psi})w' \qquad (1.120)$$

Element load vector is given by

$$[Q_F] = \int_0^L [H]^T \{a_{1x}a_{1\psi} + a_{2x}a_{2\psi} + a_{3x}a_{3\psi} + a_{4x}a_{4\psi} + a_{5x}a_{5\psi} + a_{6x}a_{6\psi} + a_{7x}a_{7\psi}$$
$$+ (b_{1x}b_{1\psi} + b_{2x}b_{2\psi})\dot{w} + (c_{1x}c_{1\psi} + c_{2x}c_{2\psi})w'\}dx$$
(2.54)

Here, $w = [H][q]$ and $\dot{w} = \frac{\partial w}{\partial \psi}$.

We write Eq. (2.54) as

$$[Q_F] = a_{1\psi}[Q_{a1}] + a_{2\psi}[Q_{a2}] + a_{3\psi}[Q_{a3}] + a_{4\psi}[Q_{a4}] + a_{5\psi}[Q_{a5}] + a_{6\psi}[Q_{a6}] + a_{7\psi}[Q_{a7}]$$
$$+ (b_{1\psi}[C_{a1}] + b_{2\psi}[C_{a2}])[\dot{q}] + (c_{1\psi}[D_{a1}] + c_{2\psi}[D_{a2}])[q]$$
(2.55)

where

$$[Q_{a1}] = \int_0^L a_{1x}[H]^T dx, \quad [Q_{a2}] = \int_0^L a_{2x}[H]^T dx, \quad [Q_{a3}] = \int_0^L a_{3x}[H]^T dx, \quad [Q_{a4}]$$
$$= \int_0^L a_{4x}[H]^T dx,$$

$$[Q_{a5}] = \int_0^L a_{5x}[H]^T dx, \quad [Q_{a6}] = \int_0^L a_{6x}[H]^T dx, \quad [Q_{a7}] = \int_0^L a_{7x}[H]^T dx,$$

$$[C_{a1}] = \int_0^L b_{1x}[H]^T[H] dx, \quad [C_{a2}] = \int_0^L b_{2x}[H]^T[H] dx,$$

$$[D_{a1}] = \int_0^L c_{1x}[H]^T[H'] dx, \text{ and } [D_{a2}] = \int_0^L c_{2x}[H]^T[H'] dx.$$

2.14 FEM Formulation of Aerodynamic Force for Rotor Problem

From Eqs. (2.52) and (2.55), we write

$$\Omega^2[M][\ddot{q}] + [K][q] = a_{1\psi}[Q_{a1}] + a_{2\psi}[Q_{a2}] + a_{3\psi}[Q_{a3}] + a_{4\psi}[Q_{a4}] + a_{5\psi}[Q_{a5}] \\ + a_{6\psi}[Q_{a6}] + a_{7\psi}[Q_{a7}] + (b_{1\psi}[C_{a1}] + b_{2\psi}[C_{a2}])[\dot{q}] + (c_{1\psi}[D_{a1}] \\ + c_{2\psi}[D_{a2}])[q] \tag{2.56}$$

We can see the presence of term involving $[q]$ and $[\dot{q}]$ on the right-hand side of this equation. These are motion-dependent forces. The presence of three forces changes the problem from a structural dynamic problem to an aeroelastic problem.

$$\Omega^2[M][\ddot{q}] + [K][q] = [Q] + [C][\dot{q}] + [D][q] \tag{2.57}$$

where

$$[Q] = a_{1\psi}[Q_{a1}] + a_{2\psi}[Q_{a2}] + a_{3\psi}[Q_{a3}] + a_{4\psi}[Q_{a4}] + a_{5\psi}[Q_{a5}] + a_{6\psi}[Q_{a6}] + a_{7\psi}[Q_{a7}]$$

$$[C] = b_{1\psi}[C_{a1}] + b_{2\psi}[C_{a2}]$$

$$[D] = c_{1\psi}[D_{a1}] + c_{2\psi}[D_{a2}]$$

After transformation of coordinate, we write Eq. (2.57) as

$$\Omega^2[M][\phi][\ddot{\zeta}] + [K][\phi][\zeta] = [Q] + [C][\phi][\dot{\zeta}] + [D][\phi][\zeta] \tag{2.58}$$

where $[q] = [\phi][\zeta]$, and $[\phi]$ being the eigenvectors. We can then write

$$\Omega^2[\phi]^T[M][\phi][\ddot{\zeta}] + [\phi]^T[K][\phi][\zeta] = [\phi]^T[Q] + [\phi]^T[C][\phi][\dot{\zeta}] + [\phi]^T[D][\phi][\zeta] \tag{2.59}$$

or

$$\Omega^2[M_1][\ddot{\zeta}] + [K_1][\zeta] = [Q_1] + [C_1][\dot{\zeta}] + [D_1][\zeta] \tag{2.60}$$

where $[M_1] = [\phi]^T[M][\phi], [K_1] = [\phi]^T[K][\phi], [Q_1] = [\phi]^T[Q], [C_1] = [\phi]^T[C][\phi],$ and $[D_1] = [\phi]^T[D][\phi]$. We notice that Eq. (2.60) is an ordinary equation having periodic coefficients and motion-dependent forcing.

We write Eq. (2.60) as

$$[A(\psi)][\ddot{\zeta}] + [B(\psi)][\dot{\zeta}] + [C(\psi)][\zeta] = [D(\psi)] \tag{2.61}$$

where $[A], [B], [C]$, and $[D]$ contain periodic functions. Thus, all the motion-dependent forces are moved to the left-hand side. This is important for solving the equation. The motion-dependent forces change the stiffness and damping terms and thus the behavior of the system. At this point, the spatial coordinate has been removed from the equation. The resulting set of ordinary differential equations now need to be solved. Chapter 3 will address the solutions of the rotor dynamics problem for the time response.

Chapter 3
Finite Element in Time

3.1 Introduction

In this chapter, finite element in time is explained with the help of examples and coupled differential equations are solved with the periodic conditions. Finite difference method (Runge–Kutta fourth order) is explained as well. Note that the helicopter blade equations are periodic differential equations.

Finite element in time is based on the weak form of Hamilton's principle

$$\int_{t_i}^{t_f} \delta L \, dt + \int_{t_i}^{t_f} \delta q^T Q \, dt = \delta q^T p \big|_{t_i}^{t_f} \tag{3.1}$$

where L is the Lagrangian of the system, and p is the set of generalized momenta. These concepts are introduced in [1].

Here, p-version of finite element in time is formulated using the continuous Galerkin's method (velocity and displacement both are continuous on the nodal boundaries).

We write Eq. (2.60) from the previous chapter

$$\Omega^2 [M_1]_1 [\ddot{\zeta}] + [K_1][\zeta] = [Q_1] + [C_1][\dot{\zeta}] + [D_1][\zeta] \tag{2.60}$$

Second-order Eq. (2.60) can be written as two first-order Eqs. (3.2) and (3.3)

$$\Omega^2 [M_1][\dot{P}] - [C_1][P] + ([K_1] - [D_1]_1)[\zeta] = [Q_1] \tag{3.2}$$

$$[P] = [\dot{\zeta}] \tag{3.3}$$

Weak formulation of Eqs. (3.2) and (3.3) in time gives Eqs. (3.4) and (3.5)

$$\int \delta W_1 \{\Omega^2[M_1][\dot{P}] - [C_1][P] + ([K_1] - [D_1])[\zeta] - [Q_1]\} d\psi = 0 \quad (3.4)$$

$$\int \delta W_2 \{[P] - [\dot{\zeta}]\} d\psi = 0 \quad (3.5)$$

Note: Equations (3.4) and (3.5) are the matrix form of the equations which are coupled; it contains a number of equations, so each equation should be formulated as weak form. For better understanding, see the example of coupled differential equations given later in this chapter.

3.2 Selection of Shape Function in Time

Assume the displacement and the velocity as a function of ψ (azimuthal angle). If the number of nodes is n, then degree of polynomial will be $n - 1$.

Example Take a case of 4 nodes $(0, 2\pi/3, 4\pi/3, 2\pi)$, where degree of polynomial is 3.

Here, u is the approximation for the displacement and v is the approximation for the velocity

$$u = a_1 + a_2\psi + a_3\psi^2 + a_4\psi^3 \quad (3.6)$$

$$v = b_1 + b_2\psi + b_3\psi^2 + b_4\psi^3 \quad (3.7)$$

Values of the displacements at the different nodal points are given by

$$u(0) = a_1$$

$$u(2\pi/3) = a_1 + 2\pi a_2/3 + 4\pi^2 a_3/9 + 8\pi^3 a_4/27$$

$$u(4\pi/3) = a_1 + 4\pi a_2/3 + 16\pi^2 a_3/9 + 64\pi^3 a_4$$

$$u(2\pi) = a_1 + 2\pi a_2 + 4\pi^2 a_3 + 8\pi^3 a_4$$

Above equations can be written in a matrix form

$$\begin{bmatrix} u(0) \\ u(2\pi/3) \\ u(4\pi/3) \\ u(2\pi) \end{bmatrix} = \begin{bmatrix} 1 & 0 & 0 & 0 \\ 1 & 2\pi/3 & 4\pi^2/9 & 8\pi^3/27 \\ 1 & 4\pi/3 & 16\pi^2/9 & 64\pi^3/27 \\ 1 & 2\pi & 4\pi^2 & 8\pi^3 \end{bmatrix} \begin{bmatrix} a_1 \\ a_2 \\ a_3 \\ a_4 \end{bmatrix} \quad (3.8)$$

3.2 Selection of Shape Function in Time

or

$$\begin{bmatrix} a_1 \\ a_2 \\ a_3 \\ a_4 \end{bmatrix} = \begin{bmatrix} 1 & 0 & 0 & 0 \\ -11/4\pi & 9/2\pi & -9/4\pi & 1/2\pi \\ 9/4\pi^2 & -45/8\pi^2 & 9/2\pi^2 & -9/8\pi^2 \\ -9/16\pi^3 & 27/16\pi3 & -27/16\pi^3 & 9/16\pi^3 \end{bmatrix} \begin{bmatrix} u(0) \\ u(2\pi/3) \\ u(4\pi/3) \\ u(2\pi) \end{bmatrix} \quad (3.9)$$

or

$$u = \begin{bmatrix} 1 & \psi & \psi^2 & \psi^3 \end{bmatrix} \begin{bmatrix} 1 & 0 & 0 & 0 \\ -11/4\pi & 9/2\pi & -9/4\pi & 1/2\pi \\ 9/4\pi^2 & -45/8\pi^2 & 9/2\pi^2 & -9/8\pi^2 \\ -9/16\pi^3 & 27/16\pi3 & -27/16\pi^3 & 9/16\pi^3 \end{bmatrix} \begin{bmatrix} u(0) \\ u(2\pi/3) \\ u(4\pi/3) \\ u(2\pi) \end{bmatrix}$$
(3.10)

or

$$u = N_1 u(0) + N_2 u(2\pi/3) + N_3 u(4\pi/3) + N_4 u(2\pi) \quad (3.11)$$

Similarly, formulation for the velocity is given by

$$v = N_1 v(0) + N_2 v(2\pi/3) + N_3 v(4\pi/3) + N_4 v(2\pi) \quad (3.12)$$

where N_1, N_2, N_3, and N_4 are the shape functions in time.

$$N_1 = 9\psi^2/4\pi^2 - 11\psi/4\pi - 9\psi^3/16\pi^3 + 1$$

$$N_2 = 9\psi/2\pi - 45\psi^2/8\pi^2 + 27\psi^3/16\pi^3$$

$$N_3 = 9\psi^2/2\pi^2 - 9\psi/4\pi - 27\psi^3/16\pi^3$$

$$N_4 = \psi/2\pi - 9\psi^2/8\pi^2 + 9\psi^3/16\pi^3$$

$$u = \begin{bmatrix} N_1 & N_2 & N_3 & N_4 \end{bmatrix} \begin{bmatrix} u_1 \\ u_2 \\ u_3 \\ u_4 \end{bmatrix} \quad (3.13)$$

or

$$u = [N][u] \quad (3.14)$$

3.3 Finite Element in Time Example

To illustrate the FEM in time formulation and concepts, we consider an example of an ordinary differential equation with periodic forcing.

$$\ddot{u} + 2\dot{u} + 3u = f(\psi) \tag{3.15}$$

where

$$\begin{aligned} f(\psi) = &\sin(\psi) + \cos(\psi) + \sin(2\psi) + \cos(2\psi) + \sin(3\psi) + \cos(3\psi) \\ &+ \sin(4\psi) + \cos(4\psi) + \sin(5\psi) + \cos(5\psi) \end{aligned}$$

and $\dot{u} = v = \frac{du}{d\psi}$.

Here, we are interested in finding the steady-state part of solution, so we assume the particular solution as

$$\begin{aligned} u_p = &a_1 \sin(\psi) + a_2 \cos(\psi) + a_3 \sin(2\psi) + a_4 \cos(2\psi) + a_5 \sin(3\psi) \\ &+ a_6 \cos(3\psi) + a_7 \sin(4\psi) + a_8 \cos(4\psi) + a_9 \sin(5\psi) + a_{10} \cos(5\psi) \end{aligned} \tag{3.16}$$

Substituting the assumed particular solution in Eq. (3.15), we get

$$a_1 = \frac{1}{2}, a_2 = 0, a_3 = \frac{3}{17}, a_4 = \frac{-5}{17}, a_5 = 0, a_6 = \frac{-1}{6},$$
$$a_7 = \frac{-5}{233}, a_8 = \frac{-21}{233}, a_9 = \frac{-3}{146}, \text{ and } a_{10} = \frac{-4}{73}.$$

The particular solution of Eq. (3.15) is given by

$$\begin{aligned} u_p = &\frac{1}{2}\sin(\psi) + \frac{3}{17}\sin(2\psi) + \frac{-5}{17}\cos(2\psi) + \frac{-1}{6}\cos(3\psi) + \frac{-5}{233}\sin(4\psi) + \frac{-21}{233}\cos(4\psi) \\ &+ \frac{-3}{146}\sin(5\psi) + \frac{-4}{73}\cos(5\psi) \end{aligned}$$
$$\tag{3.17}$$

$$\begin{aligned} v_p = &\frac{-1}{2}\cos(\psi) + \frac{6}{17}\cos(2\psi) + \frac{10}{17}\sin(2\psi) + \frac{1}{2}\sin(3\psi) + \frac{-20}{233}\cos(4\psi) + \frac{84}{233}\sin(4\psi) \\ &+ \frac{-15}{146}\cos(5\psi) + \frac{20}{73}\sin(5\psi) \end{aligned}$$
$$\tag{3.18}$$

3.3 Finite Element in Time Example

The homogenous part of Eq. (3.15) is given by

$$\ddot{u} + 2\dot{u} + 3u = 0 \tag{3.19}$$

We assume the homogenous solution to be

$$u_h = Ae^{s\psi} \tag{3.20}$$

Substituting the assumed solution in Eq. (3.19), we get

$$s = -1 \pm \sqrt{2}i$$

Substituting value of s in the assumed homogenous solution, we get

$$u_h = A_1 e^{(-1+\sqrt{2}i)\psi} + A_2 e^{(-1-\sqrt{2}i)\psi} \tag{3.21}$$

$$v_h = (-1+\sqrt{2}i)A_1 e^{(-1+\sqrt{2}i)\psi} + (-1-\sqrt{2}i)A_2 e^{(-1-\sqrt{2}i)\psi} \tag{3.22}$$

Complete solution of Eq. (3.15) is given by

$$u = u_h + u_p$$

$$\begin{aligned}
u = &\, A_1 e^{(-1+\sqrt{2}i)\psi} + A_2 e^{(-1-\sqrt{2}i)\psi} \frac{1}{2}\sin(\psi) \\
&+ \frac{3}{17}\sin(2\psi) + \frac{-5}{17}\cos(2\psi) + \frac{-1}{6}\cos(3\psi) \\
&+ \frac{-5}{233}\sin(4\psi) + \frac{-21}{233}\cos(4\psi) + \frac{-3}{146}\sin(5\psi) + \frac{-4}{73}\cos(5\psi)
\end{aligned} \tag{3.23}$$

$$v = v_h + v_p$$

$$\begin{aligned}
v = &\, (-1+\sqrt{2}i)A_1 e^{(-1+\sqrt{2}i)\psi} + (-1-\sqrt{2}i)A_2 e^{(-1-\sqrt{2}i)\psi} \\
&+ \frac{-1}{2}\cos(\psi) + \frac{6}{17}\cos(2\psi) + \frac{10}{17}\sin(2\psi) + \frac{1}{2}\sin(3\psi) \\
&+ \frac{-20}{233}\cos(4\psi) + \frac{84}{233}\sin(4\psi) + \frac{-15}{146}\cos(5\psi) + \frac{20}{73}\sin(5\psi)
\end{aligned} \tag{3.24}$$

For finite element in time formulation, we write Eq. (3.15) as two first-order equations

$$\dot{v} + 2v + 3u = f(\psi) \tag{3.25}$$

$$v = \dot{u} \tag{3.26}$$

We write Eqs. (3.25) and (3.26) in weak form

$$\int_0^{2\pi} \delta W_1(\dot{v} + 2v + 3u - f(\psi))d\psi = 0 \tag{3.27}$$

$$\int_0^{2\pi} \delta W_2(v - \dot{u})d\psi = 0 \tag{3.28}$$

Writing $u = [N][u], v = [N][v], \delta W_1 = [\delta W_1]^T[N]^T$ and $\delta W_2 = [\delta W_2]^T[N]^T$, we get equations

$$\int_0^{2\pi} [\delta W_1]^T[N]^T([\dot{N}][v] + 2[N][v] + 3[N][u] - f(\psi))d\psi = 0 \tag{3.29}$$

$$\int_0^{2\pi} [\delta W_2]^T[N]^T([N][v] - [\dot{N}][u])d\psi = 0 \tag{3.30}$$

In a matrix form, we can write Eqs. (3.29) and (3.30) as

$$\begin{bmatrix} \int_0^{2\pi} ([N]^T[\dot{N}] + 2[N]^T[N])d\psi & \int_0^{2\pi} (3[N]^T[N])d\psi \\ \int_0^{2\pi} ([N]^T[N])d\psi & -\int_0^{2\pi} ([N]^T[\dot{N}])d\psi \end{bmatrix} \begin{bmatrix} [u] \\ [v] \end{bmatrix} = \begin{bmatrix} \int_0^{2\pi} [N]^T f(\psi)d\psi \\ 0 \end{bmatrix}$$

(3.31)

From Eq. (3.31), we get the displacement and velocity at nodal points. Here, we can get the solution with periodic conditions as well as with initial conditions. Periodic conditions give us the steady-state solution, and with initial conditions, we get the transient part as well.

Here, periodic conditions are $u(0) = u(2\pi)$ and $v(0) = v(2\pi)$, and initial conditions are $u(0) = 0$ and $v(0) = 0$.

Figure 3.1a, b shows the solution using periodic conditions. Here, element length is $\pi/8$, number of elements are 16, and number of nodes within the element are 6. Figure 3.1c, d shows the solution using initial conditions. Here, element length is $\pi/4$, number of elements are 40, and number of nodes within the element are 6.

3.3 Finite Element in Time Example

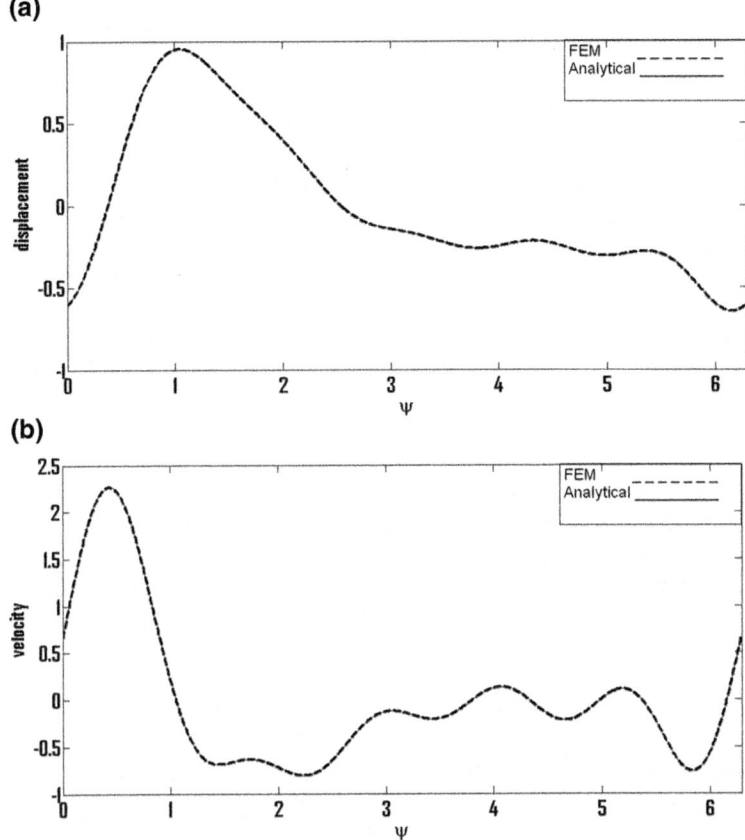

Fig. 3.1 **a** Finite element in time with periodic conditions (displacement), **b** finite element in time with periodic conditions (velocity), **c** finite element in time with initial conditions (displacement), **d** finite element in time with initial conditions (velocity)

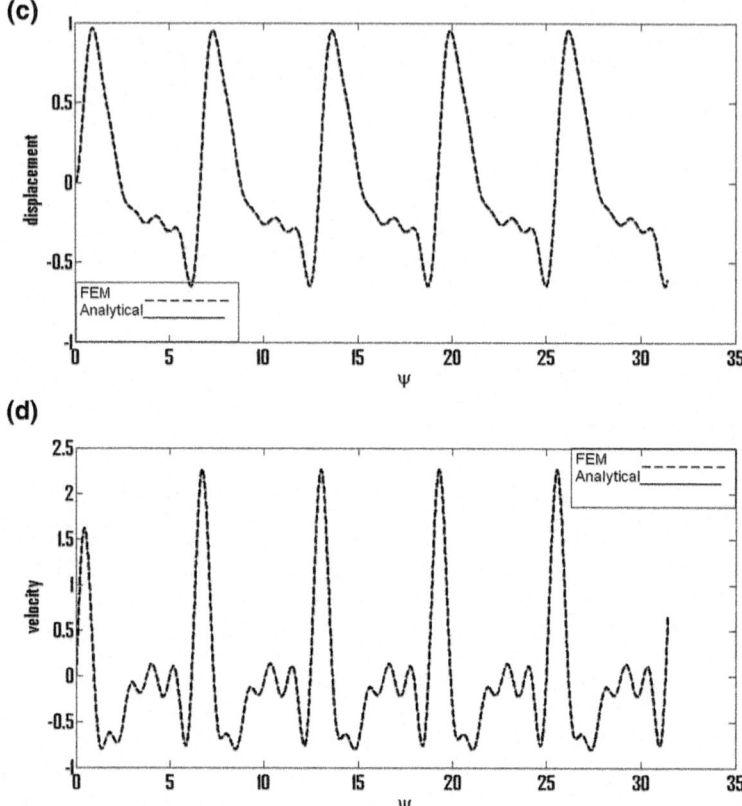

Fig. 3.1 (continued)

3.4 Solution of Coupled Differential Equations with Finite Element in Time

In case of coupled differential equations, solution may not converge with initial conditions so we use periodic conditions.

Equations (3.4) and (3.5) are coupled differential equations. We consider an example of coupled equations to illustrate the method.

$$\begin{bmatrix} 3 & 0 \\ 0 & 4 \end{bmatrix} \begin{bmatrix} \ddot{\zeta}_1 \\ \ddot{\zeta}_2 \end{bmatrix} + \begin{bmatrix} 5 & 0 \\ 0 & 9 \end{bmatrix} \begin{bmatrix} \zeta_1 \\ \zeta_2 \end{bmatrix} = \begin{bmatrix} 1 & 2 \\ 3 & 4 \end{bmatrix} \begin{bmatrix} \dot{\zeta}_1 \\ \dot{\zeta}_2 \end{bmatrix} + \begin{bmatrix} 4 & 9 \\ 6 & 7 \end{bmatrix} \begin{bmatrix} \zeta_1 \\ \zeta_2 \end{bmatrix} + \begin{bmatrix} f(\psi) \\ f(\psi) \end{bmatrix}$$
(3.32)

where $f(\psi) = 1 + \sin(\psi) + \cos(\psi) + \sin(2\psi) + \cos(2\psi)$.

3.4 Solution of Coupled Differential Equations with Finite Element in Time

Analytical solution of coupled differential Eq. (3.32) is given by

$$\zeta_1 = \frac{-11}{52} + \frac{21}{3145}\cos(\psi) - \frac{577}{3145}\sin(\psi) - \frac{-467}{6554}\cos(2\psi) - \frac{225}{6554}\sin(2\psi) \quad (3.33)$$

$$\zeta_2 = \frac{-7}{52} - \frac{46}{629}\cos(\psi) - \frac{54}{629}\sin(\psi) + \frac{117}{6554}\cos(2\psi) - \frac{505}{6554}\sin(2\psi) \quad (3.34)$$

For finite element formulation, we write two coupled differential Eqs. (3.35) and (3.36) of second order into four differential Eqs. (3.37)–(3.40) of first order. First, we move the motion-dependent term in Eq. (3.32) to the left-hand side.

$$3\ddot{\zeta}_1 - \dot{\zeta}_1 - 2\dot{\zeta}_2 + \zeta_1 - 9\zeta_2 = f(\psi) \quad (3.35)$$

$$4\ddot{\zeta}_1 - 3\dot{\zeta}_1 - 4\dot{\zeta}_2 - 6\zeta_1 + 2\zeta_2 = f(\psi) \quad (3.36)$$

$$3\dot{p}_1 - p_1 - 2p_2 + \zeta_1 - 9\zeta_2 = f(\psi) \quad (3.37)$$

$$4\dot{p}_2 - 3p_1 - 4p_2 - 6\zeta_1 + 2\zeta_2 = f(\psi) \quad (3.38)$$

$$p_1 = \dot{\zeta}_1 \quad (3.39)$$

$$p_2 = \dot{\zeta}_2 \quad (3.40)$$

Weak formulation gives us

$$\int_0^{2\pi} \delta W_1(3\dot{p}_1 - p_1 - 2p_2 + \zeta_1 - 9\zeta_2 - f(\psi))d\psi = 0 \quad (3.41)$$

$$\int_0^{2\pi} \delta W_2(4\dot{p}_2 - 3p_1 - 4p_2 - 6\zeta_1 + 2\zeta_2 - f(\psi))d\psi = 0 \quad (3.42)$$

$$\int_0^{2\pi} \delta W_3(p_1 - \dot{\zeta}_1)d\psi = 0 \quad (3.43)$$

$$\int_0^{2\pi} \delta W_4(p_2 - \dot{\zeta}_2)d\psi \quad (3.44)$$

Writing, $p_1 = [N][p_1], p_2 = [N][p_2], \zeta_1 = [N][\zeta_1], \zeta_2 = [N][\zeta_2]$, $\delta W_1 = [\delta W_1]^T[N]^T$, $\delta W_2 = [\delta W_2]^T[N]^T$, $\delta W_3 = [\delta W_3]^T[N]^T$, and $\delta W_4 = [\delta W_4]^T[N]^T$ and writing Eqs. (3.41)–(3.44) in a matrix form, we get

$$\begin{bmatrix} \int_0^{2\pi}(3[N]^T[\dot{N}]-[N]^T[N])d\psi & \int_0^{2\pi}-2[N]^T[N]d\psi & \int_0^{2\pi}[N]^T[N]d\psi & \int_0^{2\pi}-9[N]^T[N]d\psi \\ \int_0^{2\pi}-3[N]^T[N]d\psi & \int_0^{2\pi}(4[N]^T[\dot{N}]-4[N]^T[N])d\psi & \int_0^{2\pi}-6[N]^T[N]d\psi & \int_0^{2\pi}2[N]^T[N]d\psi \\ \int_0^{2\pi}[N]^T[N]d\psi & 0 & \int_0^{2\pi}-[N]^T[\dot{N}]d\psi & 0 \\ 0 & \int_0^{2\pi}[N]^T[N]d\psi & 0 & \int_0^{2\pi}-[N]^T[\dot{N}]d\psi \end{bmatrix}\begin{bmatrix}[p_1]\\[p_2]\\[\zeta_1]\\[\zeta_2]\end{bmatrix}$$

$$=\begin{bmatrix}\int_0^{2\pi}[N]^T f(\psi)d\psi \\ \int_0^{2\pi}[N]^T f(\psi)d\psi \\ 0 \\ 0\end{bmatrix}$$

(3.45)

Here, each entry in the matrix is also a matrix, and matrix size depends on the number of nodes selected. For example, for 2 nodes, each entry will be a 2×2 matrix and $[p_1]$, $[p_2]$, $[\zeta_1]$, and $[\zeta_2]$ will be a 2×1 vector.

Here, periodic conditions are used, element length is 2π number of element is 1, and number of nodes within the element are 17. Results match well with the analytical solutions as can be seen from Fig. 3.2a, b.

3.5 Enforcing Periodicity in the System

Suppose, we take only one time element with four nodes $0, 2\pi/3, 4\pi/3$, and 2π. Here, u_1, u_2, u_3, and u_4 are the displacements at the four nodes, and v_1, v_2, v_3, and v_4 are the velocities at the four nodes, respectively.

We know $u(0) = u(2\pi) \Rightarrow u_1 = u_4$ & $v(0) = v(2\pi) \Rightarrow v_1 = v_4$ and $v(0) = v(2\pi) \Rightarrow v_1 = v_4$.

$$\begin{bmatrix} a_1 & a_2 & a_3 & a_4 & b_1 & b_2 & b_3 & b_4 \\ a_5 & a_6 & a_7 & a_8 & b_5 & b_6 & b_7 & b_8 \\ a_9 & a_{10} & a_{11} & a_{12} & b_9 & b_{10} & b_{11} & b_{12} \\ a_{13} & a_{14} & a_{15} & a_{16} & b_{13} & b_{14} & b_{15} & b_{16} \\ c_1 & c_2 & c_3 & c_4 & d_1 & d_2 & d_3 & d_4 \\ c_5 & c_6 & c_7 & c_8 & d_5 & d_6 & d_7 & d_8 \\ c_9 & c_{10} & c_{11} & c_{12} & d_9 & d_{10} & d_{11} & d_{12} \\ c_{13} & c_{14} & c_{15} & c_{16} & d_{13} & d_{14} & d_{15} & d_{16} \end{bmatrix}\begin{bmatrix}v_1\\v_2\\v_3\\v_4\\u_1\\u_2\\u_3\\u_4\end{bmatrix} = \begin{bmatrix}f_1\\f_2\\f_3\\f_4\\f_5\\f_6\\f_7\\f_8\end{bmatrix} \quad (3.46)$$

Enforcing periodicity will convert Eq. (3.46) to Eq. (3.47).

3.5 Enforcing Periodicity in the System

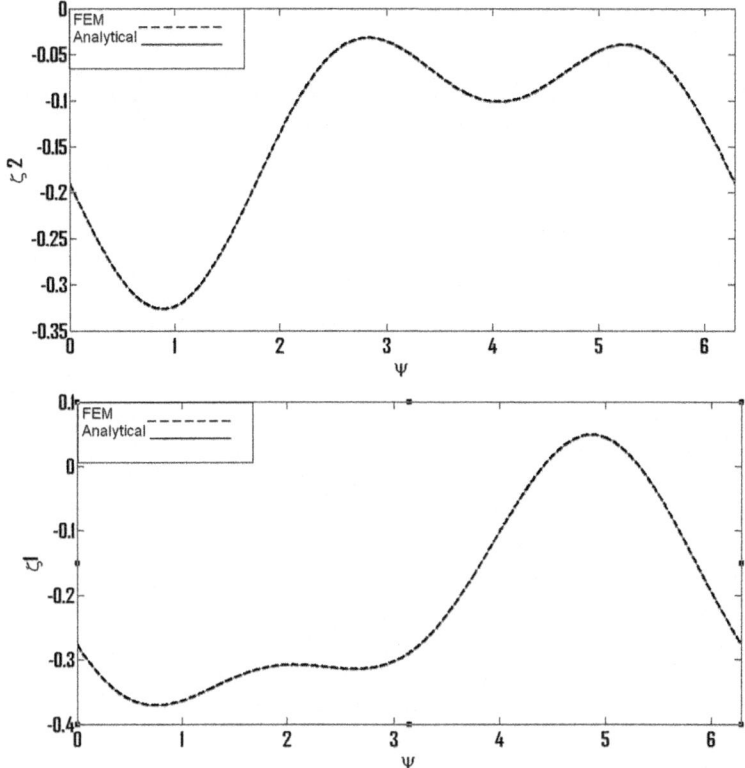

Fig. 3.2 Finite element in time for coupled differential equations

$$\begin{bmatrix} a_2+a_{14} & a_3+a_{15} & a_1+a_4+a_{13}+a_{16} & b_2+b_{14} & b_3+b_{15} & b_1+b_4+b_{13}+b_{16} \\ a_6 & a_7 & a_5+a_8 & b_6 & b_7 & b_5+b_8 \\ a_{10} & a_{11} & a_9+a_{12} & b_{10} & b_{11} & b_9+b_{12} \\ c_2+c_{14} & c_3+c_{15} & c_1+c_4+c_{13}+c_{16} & d_2+d_{14} & d_3+d_{15} & d_1+d_4+d_{13}+d_{16} \\ c_6 & c_7 & c_5+c_8 & d_6 & d_7 & d_5+d_8 \\ c_{10} & c_{11} & c_9+c_{12} & d_{10} & d_{11} & d_9+d_{12} \end{bmatrix} \begin{bmatrix} v_2 \\ v_3 \\ v_4 \\ u_2 \\ u_3 \\ u_4 \end{bmatrix} = \begin{bmatrix} f_1+f_4 \\ f_2 \\ f_3 \\ f_5+f_8 \\ f_6 \\ f_7 \end{bmatrix}$$
(3.47)

The p-version finite element in time permits straightforward application of the periodicity boundary condition.

3.6 Advantage of Choosing an Element from (0 to 2π), p-Version of Finite Element in Time

Here, we take two time elements: first is 0 to 2π and second is 2π to 4π.
Let $[N] = [N_1, N_2, N_3, N_4]$ (Shape functions for first element 0 to 2π)

Also $[M] = [M_1, M_2, M_3, M_4]$ (Shape function for second element 2π to 4π)
We notice that

$$\int_0^{2\pi} A(\psi)[N']^T[N]\,d\psi = \int_{2\pi}^{4\pi} A(\psi)[M']^T[M]\,d\psi \tag{3.48}$$

$A(\psi)$ is a periodic function with a period of 2π. We can use shape function of one element for all the elements if the length of the each element is 2π.

3.7 Selection of Number of Nodes

Here, we take a differential equation with the periodic forcing.

$$\ddot{u} + 2\dot{u} + 3u = \sin(\psi) + \cos(\psi) + \sin(2\psi) + \cos(2\psi) \tag{3.49}$$

where $\dot{u} = v = \frac{du}{d\psi}$.

Particular solution of Eq. (3.49) is given by

$$u_p = \frac{1}{2}\sin(\psi) + \frac{3}{17}\sin(2\psi) + \frac{-5}{17}\cos(2\psi) \tag{3.50}$$

$$v_p = \frac{-1}{2}\cos(\psi) + \frac{6}{17}\cos(2\psi) + \frac{10}{17}\sin(2\psi) \tag{3.51}$$

Equation (3.49) is solved using finite element in time with only one time element; length of the element is 0 to 2π. Periodic conditions are used to get steady-state solution. Three cases are considered here with different number of nodes. The results in Fig. 3.3a–c show excellent agreement with analytical solution.

Case1—(1 element, 6 nodes)
Case2—(1 element, 11 nodes)
Case3—(1 element, 17 nodes)

3.8 Effect of Forcing Term in Finite Element in Time

Here, we compare three Eqs. (3.51), (3.54) and (3.57), with 11 node, 1 time element in 0 to 2π. These equations model an increase in the order of forcing.

$$\ddot{u} + 2\dot{u} + 3u = \sin(\psi) + \cos(\psi) \tag{3.51}$$

where $\dot{u} = v = \frac{du}{d\psi}$.

3.8 Effect of Forcing Term in Finite Element in Time

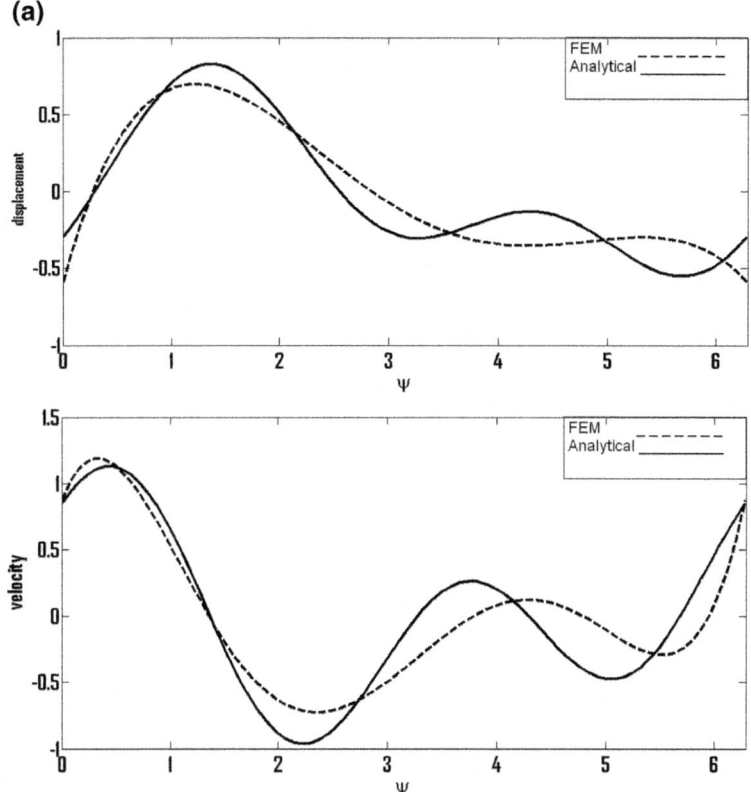

Fig. 3.3 a Selection of number of nodes (1 element, 6 nodes), **b** selection of number of nodes (1 element, 11 nodes), **c** selection of number of nodes (1 element, 17 nodes)

Particular solution of Eq. (3.51) is given by

$$u_p = \frac{1}{2}\sin(\psi) \tag{3.52}$$

$$v_p = \frac{-1}{2}\cos(\psi) \tag{3.53}$$

In Fig. 3.4a, we see that the FEM correlates well with the analytical results.

$$\ddot{u} + 2\dot{u} + 3u = \sin(\psi) + \cos(\psi) + \sin(2\psi) + \cos(2\psi) \tag{3.54}$$

where $\dot{u} = v = \frac{du}{d\psi}$.

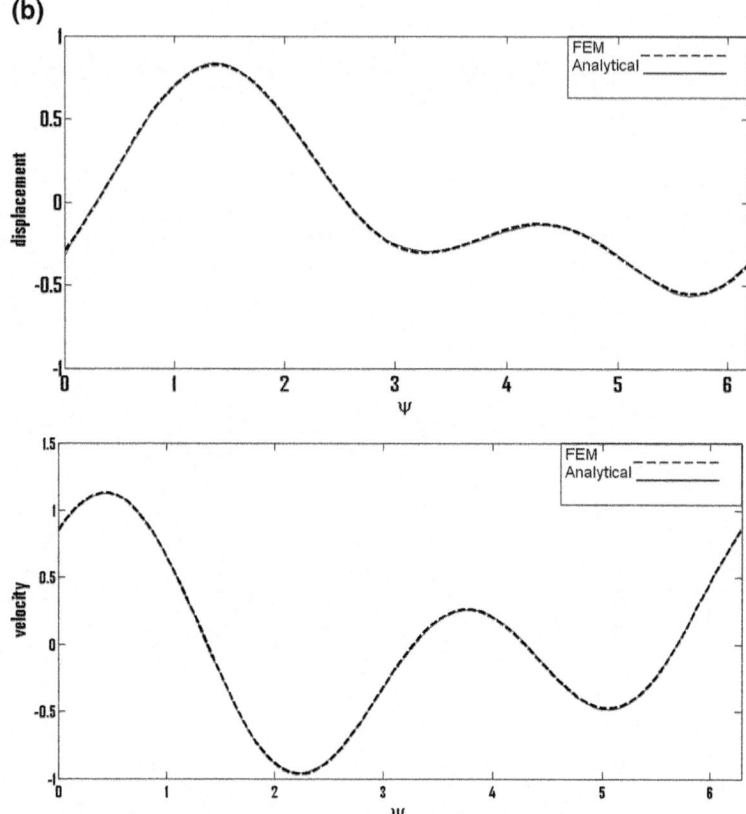

Fig. 3.3 (continued)

Particular solution of Eq. (3.54) is given by

$$u_p = \frac{1}{2}\sin(\psi) + \frac{3}{17}\sin(2\psi) + \frac{-5}{17}\cos(2\psi) \tag{3.55}$$

$$v_p = \frac{-1}{2}\cos(\psi) + \frac{6}{17}\cos(2\psi) + \frac{10}{17}\sin(2\psi) \tag{3.56}$$

Figure 3.4b also shows good agreement. However, Fig. 3.4c shows a deterioration in performance as the polynomial discretization is insufficient for the high degree of forcing.

(c)

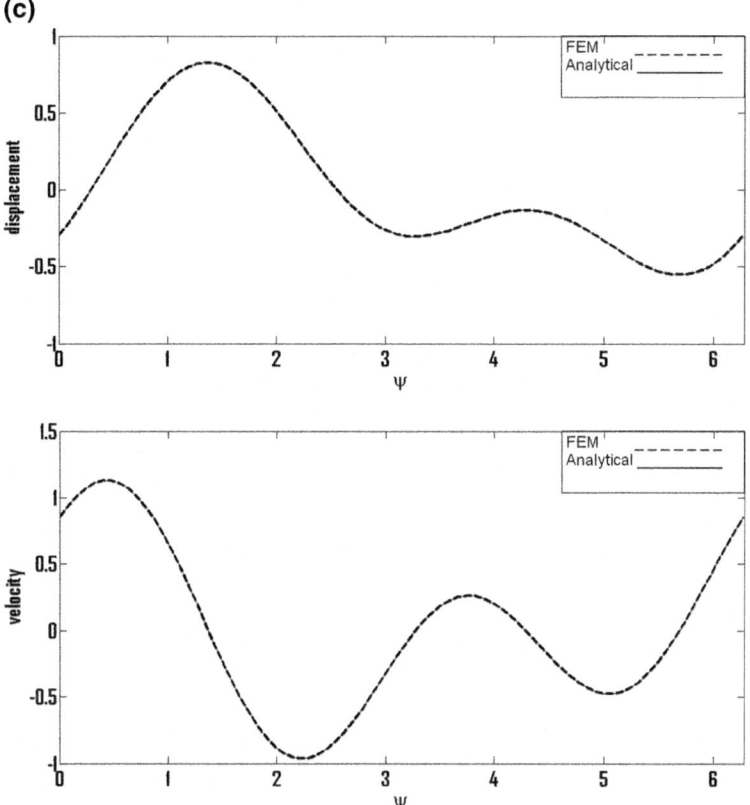

Fig. 3.3 (continued)

$$\ddot{u} + 2\dot{u} + 3u = \sin(\psi) + \cos(\psi) + \sin(2\psi) + \cos(2\psi) + \sin(3\psi) + \cos(3\psi) \tag{3.57}$$

where $\dot{u} = v = \frac{du}{d\psi}$.

Particular solution of Eq. (3.57) is given by

$$u_p = \frac{1}{2}\sin(\psi) + \frac{3}{17}\sin(2\psi) + \frac{-5}{17}\cos(2\psi) + \frac{-1}{6}\cos(3\psi) \tag{3.58}$$

$$v_p = \frac{-1}{2}\cos(\psi) + \frac{6}{17}\cos(2\psi) + \frac{10}{17}\sin(2\psi) + \frac{1}{2}\sin(3\psi) \tag{3.59}$$

Hence, we can see how the forcing affects our results. When there is a higher harmonic content in the forcing, a lower number of nodes are insufficient.

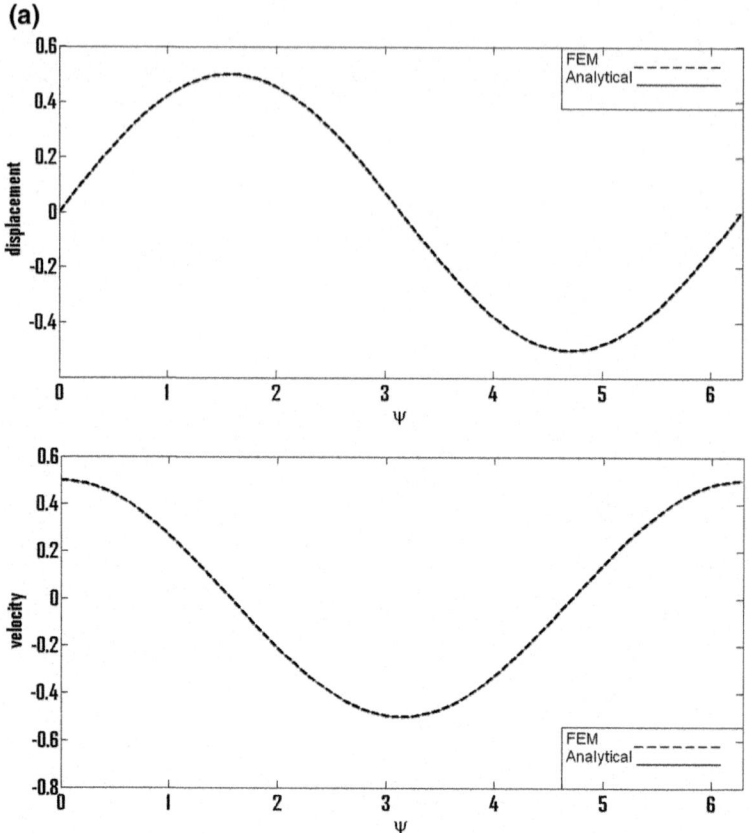

Fig. 3.4 **a** Effect of forcing (1 element, 11 nodes, $f(\psi) = \sin(\psi) + \cos(\psi)$), **b** effect of forcing (1 element, 11 nodes, $f(\psi) = \sin(\psi) + \cos(\psi) + \sin(2\psi) + \cos(2\psi)$), **c** effect of forcing (1 element, 11 nodes, $f(\psi) = \sin(\psi) + \cos(\psi) + \sin(2\psi) + \cos(2\psi) + \sin(3\psi) + \cos(3\psi)$), **d** effect of forcing (1 element, 11 nodes, $f(\psi) = \sin(\psi) + \cos(\psi) + \sin(2\psi) + \cos(2\psi) + \sin(3\psi) + \cos(3\psi)$)

Now, we solve Eq. (3.41), with 17 node, 1 time element in 0 to 2π.

We can see in Fig. 3.4b–d, Eq. (3.54) can be solved with lesser number of nodes, but for Eq. (3.57), higher number of nodes are required within the same time element. For a typical problem, the analytical solution is not known. Thus, it is a good idea to perform a convergence study for a given forcing function. The number of nodes is increased and the value of nodes fixed at the point where the response does not change due to an additional node.

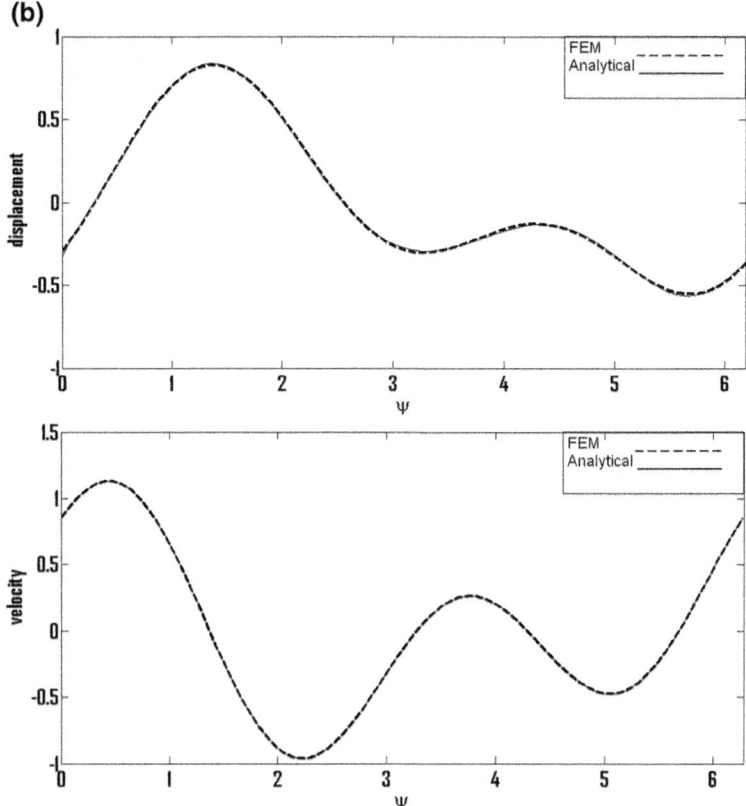

Fig. 3.4 (continued)

3.9 Finite Difference Method (Runge–Kutta Fourth Order)

We solve a periodic differential equation using Runge–Kutta fourth-order method. Such a method helps to verify the finite element in time results.

$$\ddot{u} + 2\dot{u} + 3u = f(\psi) \tag{3.15}$$

where

$$f(\psi) = \sin(\psi) + \cos(\psi) + \sin(2\psi) + \cos(2\psi) + \sin(3\psi) + \cos(3\psi) \\ + \sin(4\psi) + \cos(4\psi) + \sin(5\psi) + \cos(5\psi)$$

We write the second-order Eq. (3.15) as two first-order differential equations

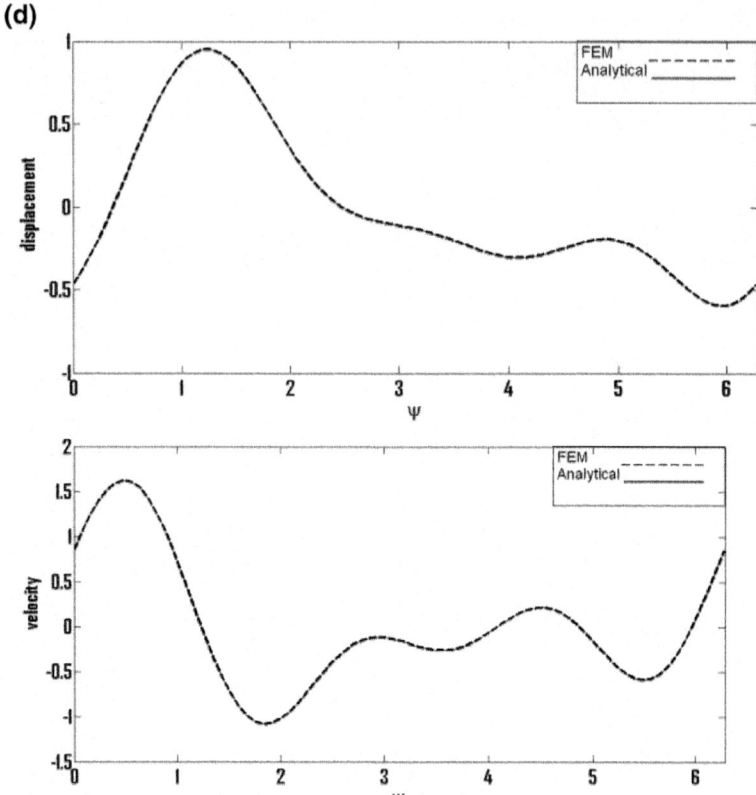

Fig. 3.4 (continued)

$$\dot{v} + 2v + 3u = f(\psi) \Rightarrow \dot{v} = -2v - 3u + f(\psi) \qquad (3.60)$$

$$v = \dot{u} \Rightarrow \dot{u} = v \qquad (3.61)$$

We write Eqs. (3.42) and (3.43) in the form of Eqs. (3.44) and (3.45) with initial conditions $u(0) = 0$, and $v(0) = 0$.

$$\frac{dv}{d\psi} = f_1(u, v, \psi), v(0) = 0 \qquad (3.62)$$

$$\frac{du}{d\psi} = f_2(u, v, \psi), u(0) = 0 \qquad (3.63)$$

We solve Eqs. (3.44) and (3.45) using Runge–Kutta method. We March in time using following equations

3.9 Finite Difference Method (Runge–Kutta Fourth Order)

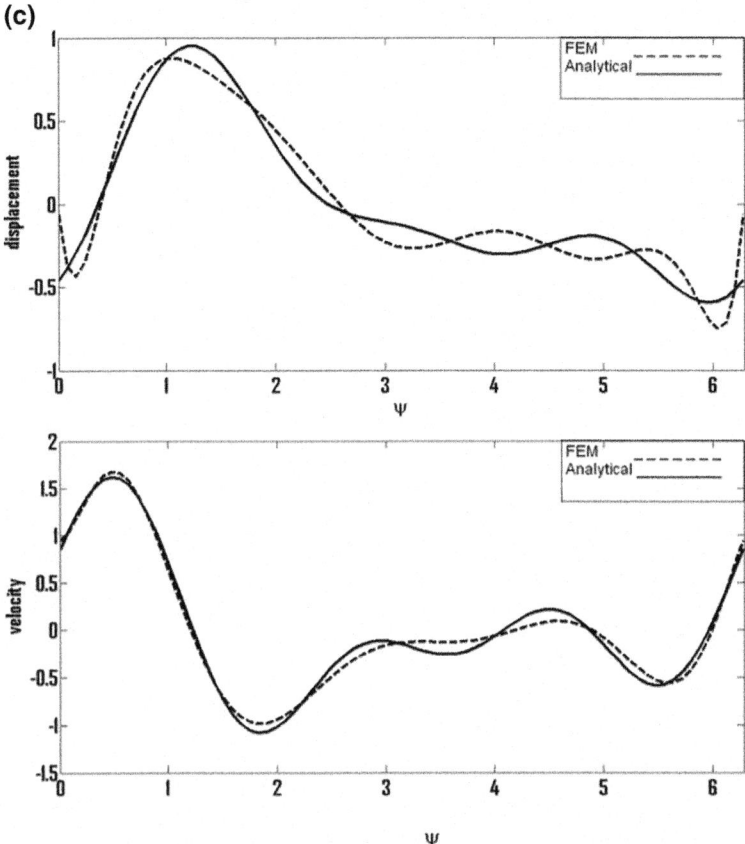

Fig. 3.4 (continued)

$$u_{i+1} = u_i + 16(ku_1 + 2ku_2 + 2ku_3 + ku_4)h \qquad (3.64)$$

$$v_{i+1} = v_i + 16(kv_1 + 2kv_2 + 2kv_3 + kv_4)h \qquad (3.65)$$

where

$$ku_1 = f_2(u_i, v_i, \psi_i)$$

$$kv_1 = f_1(u_i, v_i, \psi_i)$$

$$ku_2 = f_2\left(u_i + ku_1\frac{h}{2}, v_i + kv_1\frac{h}{2}, \psi_i + \frac{h}{2}\right)$$

$$kv_2 = f_1\left(u_i + ku_1\frac{h}{2}, v_i + kv_1\frac{h}{2}, \psi_i + \frac{h}{2}\right)$$

$$ku_3 = f_2\left(u_i + ku_2\frac{h}{2}, v_i + kv_2\frac{h}{2}, \psi_i + \frac{h}{2}\right)$$

$$kv_3 = f_1\left(u_i + ku_2\frac{h}{2}, v_i + kv_2\frac{h}{2}, \psi_i + \frac{h}{2}\right)$$

$$ku_4 = f_2(u_i + ku_3 h, v_i + kv_3 h, \psi + h)$$

$$kv_4 = f_1(u_i + ku_3 h, v_i + kv_3 h, \psi + h)$$

Result of Runge–Kutta method with a time step is $\pi/32$ is shown in Fig. 3.5a. A very similar result was obtained with finite element in time in the Sect. (3.3) for Eq. (3.15).

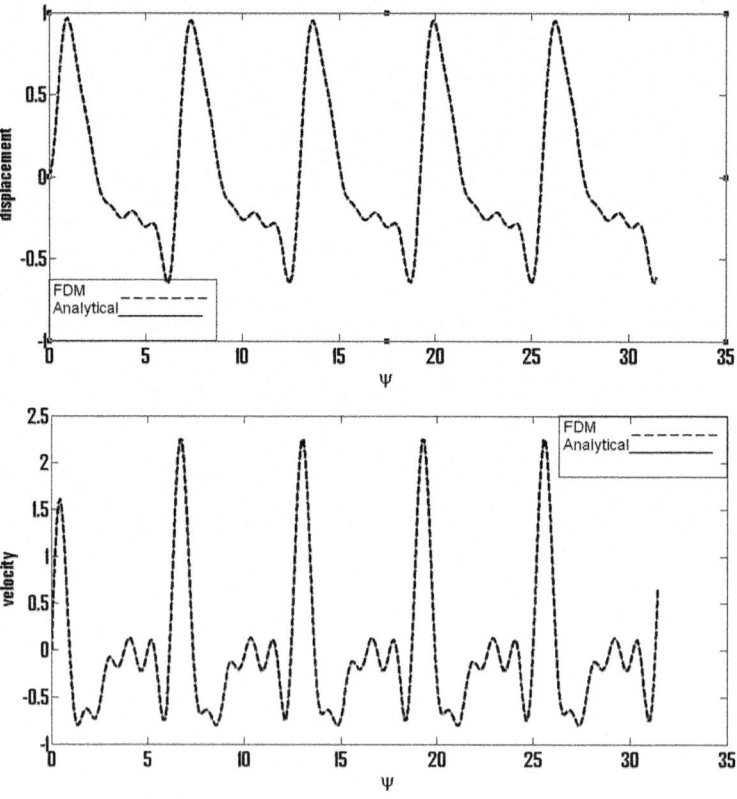

Fig. 3.5 Runge–Kutta fourth-order result

3.9 Finite Difference Method (Runge–Kutta Fourth Order)

Thus, helicopter rotor problems can be solved using the finite element in time or Runge–Kutta method. Finite element in time is useful to get the periodic response efficiently. In most comprehensive aeroelastic analysis, response, loads, and stability are calculated for steady-state condition. Therefore, finite element in time is often used. Since the problem is 1D in time, a p-version time finite element works very well and is illustrated in this book.

Chapter 4
Stability Analysis

4.1 Introduction

Stability analysis of a system includes calculation of the eigenvalues of the matrix containing the physics of the governing differential equation. If a differential equation has constant coefficients, eigenvalues can be calculated analytically. In rotor problem, we get a differential equation, which has periodic coefficients, and a numerical solution is possible. Here, Floquet theory is used to find out the eigenvalues of a periodic system. Equations (4.1) and (4.2) represent a differential equation with non-periodic coefficients and a differential equation with periodic coefficients, respectively. We will use these examples to illustrate the method used in stability analysis.

$$\ddot{u} + 2\dot{u} + 3u = \sin(\psi) + \cos(\psi) \tag{4.1}$$

$$\sin(\psi)\ddot{u} + 2\cos(\psi)\dot{u} + 3\{\sin(\psi) + \cos(\psi)\}u = \sin(\psi) + \cos(\psi) \tag{4.2}$$

4.2 Stability Analysis of Equations with Constant Coefficients

Here, we take two differential Eqs. (4.3) and (4.4) with constant coefficients.

$$\ddot{u} + 2\dot{u} + 3u = \sin(\psi) + \cos(\psi) \tag{4.3}$$

$$\ddot{u} - 2\dot{u} + 3u = \sin(\psi) + \cos(\psi) \tag{4.4}$$

For stability analysis, we write state-space representation of differential Eqs. (4.3) and (4.4)

$$[\dot{X}(t)]_{n\times 1} = [A]_{n\times n}[X(t)]_{n\times 1} + [B]_{n\times n}[u(t)]_{n\times 1} \tag{4.5}$$

For a system to be stable, the real part of all the eigenvalues of the matrix $[A]$ has to be negative. Equations (4.6) and (4.7) are state-space representation of differential Eqs. (4.3) and (4.4) respectively.

$$\begin{bmatrix} \frac{du}{dt} \\ \frac{dv}{dt} \end{bmatrix} = \begin{bmatrix} 0 & 1 \\ -3 & -2 \end{bmatrix}\begin{bmatrix} u \\ v \end{bmatrix} + \begin{bmatrix} 0 \\ \sin(\psi)+\cos(\psi) \end{bmatrix} \tag{4.6}$$

$$\begin{bmatrix} \frac{du}{dt} \\ \frac{dv}{dt} \end{bmatrix} = \begin{bmatrix} 0 & 1 \\ -3 & 2 \end{bmatrix}\begin{bmatrix} u \\ v \end{bmatrix} + \begin{bmatrix} 0 \\ \sin(\psi)+\cos(\psi) \end{bmatrix} \tag{4.7}$$

Eigenvalues of the matrix $[A]$ of differential Eq. (4.6) are $-1 \pm \sqrt{2}i$ and eigenvalues of the matrix $[A]$ of differential Eq. (4.7) are $1 \pm \sqrt{2}i$. For a system to be stable, real part of the eigenvalues has to be negative Therefore, Eq. (4.6) is a stable system, and Eq. (4.7) is an unstable system. Such systems are sometimes displayed in a root locus plot as shown in Fig. 4.1a, b.

If all of the eigenvalues lie in the left-half plane, the system is stable. If any of the eigenvalues lie on the right-half plane, the system is unstable. Eigenvalues on the y-axis show neutral stability.

If we solve Eqs. (4.6) and (4.7), we get the result shown in Fig. 4.1c, d, respectively. The problem of instability is seen here in the response of the system. Note that stability is an intrinsic characteristic of the system. Equations (4.6) and (4.7) have the same forcing but different coefficient matrices. But the coefficient matrix can completely change the nature of the system.

4.3 Stability Analysis of a Coupled Differential Equations with Constant Coefficients

We take a coupled differential equations with constant coefficients

$$[A][\ddot{\zeta}] + [B][\dot{\zeta}] + [C][\zeta] = [D] \tag{4.8}$$

or

$$\begin{bmatrix} 3 & 0 \\ 0 & 4 \end{bmatrix}\begin{bmatrix} \ddot{\zeta}_1 \\ \ddot{\zeta}_2 \end{bmatrix} + \begin{bmatrix} 5 & 0 \\ 0 & 9 \end{bmatrix}\begin{bmatrix} \dot{\zeta}_1 \\ \dot{\zeta}_2 \end{bmatrix} - \begin{bmatrix} 1 & 2 \\ 3 & 4 \end{bmatrix}\begin{bmatrix} \zeta_1 \\ \zeta_2 \end{bmatrix} = \begin{bmatrix} \sin(\psi) \\ \cos(\psi) \end{bmatrix} \tag{4.9}$$

4.3 Stability Analysis of a Coupled Differential Equations ...

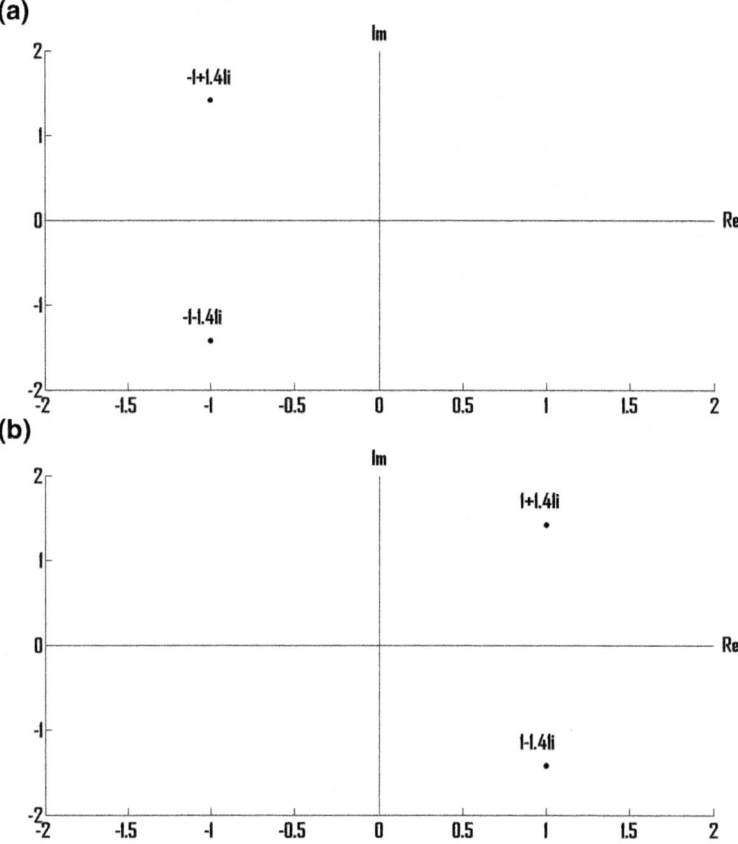

Fig. 4.1 **a** Stable system, root locus plot of differential Eq. (4.3), **b** unstable system, root locus plot of differential Eq. (4.4), **c** stable system, solution of differential Eq. (4.3), **d** unstable system, solution of differential Eq. (4.4)

where $[A] = \begin{bmatrix} 3 & 0 \\ 0 & 4 \end{bmatrix}, [B] = \begin{bmatrix} 5 & 0 \\ 0 & 9 \end{bmatrix}, [C] = -\begin{bmatrix} 1 & 2 \\ 3 & 4 \end{bmatrix}$, and $[D] = \begin{bmatrix} \sin(\psi) \\ \cos(\psi) \end{bmatrix}$

We can write state-space representation of Eq. (4.8)

$$\begin{bmatrix} [\dot{\zeta}]_{2\times1} \\ [\dot{P}]_{2\times1} \end{bmatrix} = \begin{bmatrix} [0]_{2\times2} & [I]_{2\times2} \\ -[A]^{-1}_{2\times2}[C]_{2\times2} & -[A]^{-1}_{2\times2}[B]_{2\times2} \end{bmatrix} \begin{bmatrix} \zeta \\ P \end{bmatrix} + \begin{bmatrix} [0]_{2\times2} \\ [A]^{-1}_{2\times2}[D]_{2\times2} \end{bmatrix} \quad (4.10)$$

where $[P] = [\dot{\zeta}] = \begin{bmatrix} P_1 \\ P_2 \end{bmatrix}$

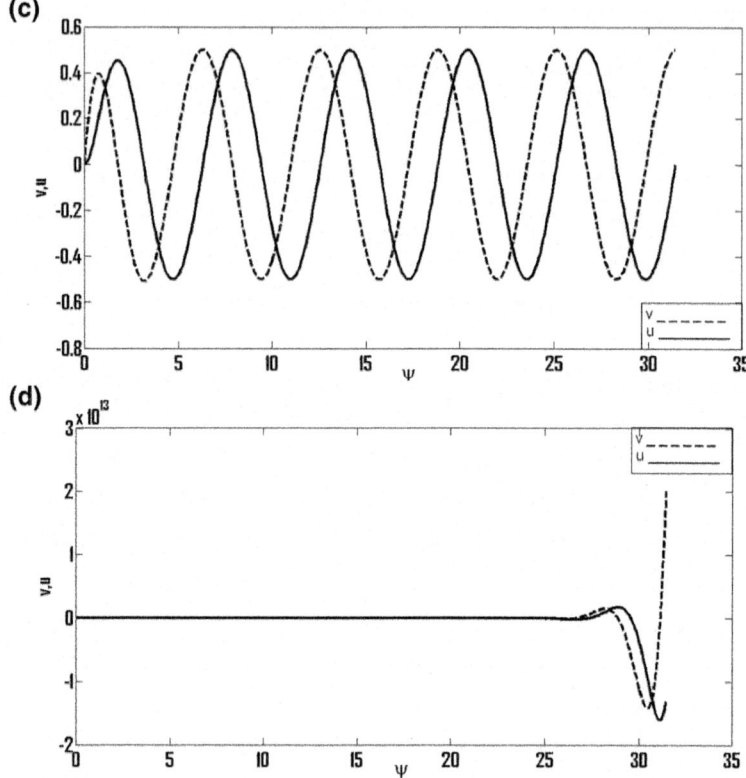

Fig. 4.1 (continued)

or

$$\begin{bmatrix} \dot{\zeta}_1 \\ \dot{\zeta}_2 \\ \dot{P}_1 \\ \dot{P}_2 \end{bmatrix} = \begin{bmatrix} 0 & 0 & 1 & 0 \\ 0 & 0 & 0 & 1 \\ 0.33 & 0.66 & -1.66 & 0 \\ 0.75 & 1 & 0 & -2.25 \end{bmatrix} \begin{bmatrix} \zeta_1 \\ \zeta_2 \\ P_1 \\ P_2 \end{bmatrix} + \begin{bmatrix} 0 \\ 0 \\ \sin(\psi)/3 \\ \cos(\psi)/4 \end{bmatrix} \quad (4.11)$$

Eigenvalues of the system are 0.5537, −0.0649, −2.6956, and −1.7033. Real part of all the eigenvalues is not negative, so system is unstable. In hover condition, the helicopter rotor equation has constant coefficients, and the methods discussed above can be used.

4.4 Stability Analysis of the Equation with Periodic Coefficients, Floquet Theory

Consider the equation

$$A(\psi)\ddot{\zeta} + B(\psi)\dot{\zeta} + C(\psi)\zeta = D(\psi) \quad (4.12)$$

To find stability solution for Eq. (4.12), we should use Floquet theory. We have the state-space representation.

$$[\dot{X}(t)] = [A(t)][X(t)] \quad (4.13)$$

Solution of Eq. (4.13) is given by [9]

$$[X(t)] = [\phi(t - t_0)][X(t)_0] \quad (4.14)$$

where $[X(t_0)]$ is initial value at $t = 0$ and $[\phi(t - t_0)]$ is a transition matrix.

Our goal is to find out the eigenvalues of the matrix $[A(t)]$. If we know the eigenvalues of a transition matrix $[\phi(t - t_0)]$, we can find the eigenvalues of the matrix $[A(t)]$.

Suppose we know that one of the eigenvalues of the transition matrix $[\phi(t - t_0)]$ is H and we want to find out what will be the eigenvalues (λ) of the matrix $[A(t)]$ after time period T; relation between the two eigenvalues is given by

$$\lambda = \frac{1}{T}\ln(H) + i\left(\frac{1}{T}\right)\tan^{-1}\left[\frac{\text{Im}(H)}{\text{Re}(H)}\right] \quad (4.15)$$

or

$$\lambda = \xi_P + iw_P \quad (4.16)$$

where $\xi_P = \frac{1}{T}\ln(H)$ represents damping rates and $\omega_P = \left(\frac{1}{T}\right)\tan^{-1}\left[\frac{\text{Im}(H)}{\text{Re}(H)}\right]$ represents frequencies. Since $\tan^{-1}\left[\frac{\text{Im}(H)}{\text{Re}(H)}\right]$ is multi-valued, there can be ambiguous answers for frequencies.

4.5 Analytical Solution with the Floquet Theory

Here, we solve a stability problem which includes periodic functions

$$\begin{bmatrix} \frac{dx}{d\psi} \\ \frac{dy}{d\psi} \end{bmatrix} = \begin{bmatrix} \sin(\psi) & 0 \\ e^{\cos(\psi)} & 0 \end{bmatrix} \begin{bmatrix} x \\ y \end{bmatrix} + \begin{bmatrix} 0 \\ F\sin(\psi) \end{bmatrix} \quad (4.17)$$

For stability analysis, we substitute $F = 0$. We get the equation in the form of $[\dot{X}(t)] = [A(t)][X(t)]$

$$\begin{bmatrix} \frac{dx}{d\psi} \\ \frac{dy}{d\psi} \end{bmatrix} = \begin{bmatrix} \sin(\psi) & 0 \\ e^{\cos(\psi)} & 0 \end{bmatrix} \begin{bmatrix} x \\ y \end{bmatrix} \quad (4.18)$$

Time period $T = 2\pi$. We take two sets of initial conditions $[X(\psi_0)] = \begin{bmatrix} x(0) \\ y(0) \end{bmatrix} = \begin{bmatrix} 1 \\ 0 \end{bmatrix}$ and $[X(\psi_0)] = \begin{bmatrix} x(0) \\ y(0) \end{bmatrix} = \begin{bmatrix} 0 \\ 1 \end{bmatrix}$ to find out the transition matrix $\phi(t)$.

Solution of Eq. (4.18) is

$$x = c_1 e^{-\cos(\psi)} \quad (4.19)$$

$$y = c_1 \psi + c_2 \quad (4.20)$$

We apply the 1st set of initial condition to Eqs. (4.19) and (4.20) to get $c_1 = e$ and $c_2 = 0$.

We apply the 2nd set of the initial condition to Eqs. (4.19) and (4.20) to get $c_1 = 0$ and $c_2 = 1$.

Suppose we want to know solution after $T = 2\pi$. Equations. (4.19) and (4.20) with the 1st set of initial conditions, after time period $T = 2\pi$, give us

$$x = 1, \; y = 2\pi e$$

Equations (4.19) and (4.20) with the 2nd set of initial conditions, after time period $T = 2\pi$, give us

$$x = 0, \; y = 1$$

We relate the solution at $T = 2\pi$, with solution at $T = 0$, in Eqs. (4.21) and (4.22)

$$\begin{bmatrix} 1 \\ 2\pi e \end{bmatrix} = \begin{bmatrix} 1 & 0 \\ 2\pi e & 1 \end{bmatrix} \begin{bmatrix} 1 \\ 0 \end{bmatrix} \quad (4.21)$$

$$\begin{bmatrix} 0 \\ 1 \end{bmatrix} = \begin{bmatrix} 1 & 0 \\ 2\pi e & 1 \end{bmatrix} \begin{bmatrix} 0 \\ 1 \end{bmatrix} \quad (4.22)$$

4.5 Analytical Solution with the Floquet Theory

Equations (4.21) and (4.22) are in state-space representation form

$$[X(2\pi)] = [\phi][X(0)] \tag{4.23}$$

From Eqs. (4.21), (4.22), and (4.23), we get

$$[\phi] = \begin{bmatrix} 1 & 0 \\ 2\pi e & 1 \end{bmatrix} \tag{4.24}$$

Eigenvalues of this transition matrix are (1, 1). From Eq. (4.15), we get the eigenvalues (0, 0) of the matrix $[A(2\pi)]$. The system is neutrally stable.

4.6 Numerical Method to Evaluate a Transition Matrix

Here, we get a transition matrix numerically. Time period T is divided into k equal intervals, and in each interval, the matrix $[A(t)]$ is replaced by the matrix $[C_k]$ with constant coefficients.

We take increment in time as

$$0 = \psi_0 < \psi_1 < \psi_2 \ldots < \psi_{k-1} < \psi_k = T,$$

where kth interval is (ψ_{k-1}, ψ_k)

$$[C_k] = \frac{1}{\Delta_k} \int_{\psi_{K-1}}^{\psi_K} A[\zeta] \mathrm{d}\zeta \tag{4.25}$$

where $\Delta_k = \psi_k - \psi_{k-1}$.

Transition matrix after time period T is given by Eq. (4.26).

$$[\phi(T, 0)] = \exp(\Delta_k [c_k]) * \exp(\Delta_{k-1} [c_{k-1}]) * \exp(\Delta_{k-2} [c_{k-2}]) \ldots \exp(\Delta_1 [c_1]) \tag{4.26}$$

This concept is easier to understand through an example.

Example Take state-space representation (4.27)

$$\begin{bmatrix} \dot{x}_1(t) \\ \dot{x}_2(t) \end{bmatrix} = \begin{bmatrix} \sin(t) & 2\cos(t) + 3 \\ \cos^3(t) & 4 \end{bmatrix} \begin{bmatrix} x_1(t) \\ x_2(t) \end{bmatrix} \tag{4.27}$$

Suppose we want to find stability after a period of 2π. We divide the period $T = 2\pi$ into 32 divisions, each division having an interval of $\pi/32$.

Here,

$$\Delta_k = \frac{\pi}{16} - 0 = \frac{2\pi}{16} - \frac{\pi}{16} = \frac{3\pi}{16} - \frac{2\pi}{16} \ldots = \frac{32\pi}{16} - \frac{31\pi}{16},$$

$$\Delta_1 = \Delta_2 = \Delta_3 \ldots \Delta_{32},$$

$$[C_1] = \frac{1}{\frac{\pi}{16}} \int_0^{\frac{\pi}{16}} A(t)dt, [C_2] = \frac{1}{\frac{2\pi}{16} - \frac{\pi}{16}} \int_{\frac{\pi}{16}}^{\frac{2\pi}{16}} A(t)dt, \ldots [C_{32}] = \frac{1}{2\pi - \frac{31\pi}{16}} \int_{\frac{31\pi}{16}}^{2\pi} A(t)dt$$

Transition matrix $\phi(2\pi, 0)$ is given by

$$[\phi(2\pi, 0)] = \exp(\Delta_{32}[C_{32}]) \cdot \exp(\Delta_{31}[C_{31}]) \ldots \exp(\Delta_1[C_1]) \tag{4.28}$$

For example (4.17), we get

$$[\phi_{\text{Analytical}}] = \begin{bmatrix} 1 & 0 \\ 2\pi e & 1 \end{bmatrix} = \begin{bmatrix} 1 & 0 \\ 17.0795 & 1 \end{bmatrix} \text{ and } [\phi_{\text{Numerical}}] = \begin{bmatrix} 1 & 0 \\ 17.0802 & 1 \end{bmatrix}$$

198 equal intervals are taken while solving numerically.

4.7 Stability Analysis for Rotor Problem

For rotor problem, we get Eq. (2.60) after the finite element in space

$$\Omega^2[M_1][\ddot{\zeta}] + [K_1][\zeta] = [Q_1] + [C_1][\dot{\zeta}] + [D_1][\zeta] \tag{2.60}$$

For stability analysis, we solve Eq. (4.29)

$$\Omega^2[M_1][\ddot{\zeta}] + [K_1][\zeta] - [C_1][\dot{\zeta}] - [D_1][\zeta] = 0 \tag{4.29}$$

We write Eq. (4.29) into two first-order differential equations

$$\Omega^2[M_1][\dot{p}] - [C_1][p] + ([K_1] - [D_1])[\zeta] = 0 \tag{4.30}$$

$$[P] = [\dot{\zeta}] \tag{4.31}$$

We write Eq. (4.30) as

$$[M_2][\dot{P}] + [C_2][P] + [K_2]\zeta = 0 \tag{4.32}$$

where $[M_2] = \Omega^2[M_1]$, $[K_2] = [K_1 - D_1]$, and $[C_2] = -[C_1]$.

4.7 Stability Analysis for Rotor Problem

We write Eqs. (4.31) and (4.32) in matrix form as

$$\begin{bmatrix} [\dot{\zeta}] \\ [\dot{P}] \end{bmatrix} = \begin{bmatrix} [0] & [I] \\ -[M_2]^{-1}[K_2] & -[M_2]^{-1}[C_2] \end{bmatrix} \begin{bmatrix} [\zeta] \\ [P] \end{bmatrix} \quad (4.33)$$

Equation (4.33) is in state-space representation form $[\dot{X}(t)] = [A(t)][X(t)]$. We can find out the eigenvalues of the matrix $[A(t)]$ numerically using the Floquet theory. Chapter 5 illustrates the numerical results for the helicopter rotor problem.

Chapter 5
Helicopter Rotor Results

The theory discussed in previous chapters is now applied to a helicopter rotor. The results are matched with published literature. A MATLAB code is given along with this book which can be used to generate results given in this chapter.

5.1 Inputs

Table 5.1 shows the inputs [3], used for the elastic rotor problem. These are typical of a four-bladed hingeless rotor. In this book, we assume a flapping elastic blade with cantilever boundary conditions.

5.2 Result 1 (Mode Shapes and Frequencies of the Rotating Beam)

We have discussed the mode shape and the frequency calculation in Sect. 2.13. The non-dimensional rotating frequencies are shown in Table 5.2 and matched with [6], for the given values of non-dimensional rotating speed s. The beam is divided into 200 finite elements with four degrees of freedom for each element. The rotation leads to an increase in the beam stiffness, and therefore in the natural frequency. We observe this stiffening effect for the first mode. For the second and higher modes, stiffening comes largely from flexure, and the effect of rotation is much less.

We draw a Campbell diagram (Fig. 5.1) and observe the stiffening effect in the rotating beam with the increase in the rotating speed, which will result increased natural frequencies. Here $\omega_1, \omega_2, \omega_3, \omega_4,$ and ω_5 are the first five natural frequencies of the rotating beam, respectively.

Table 5.1 Inputs for the elastic rotor problem

Radius	R	4.9378 m
Chord/Radius	c/R	0.08
Rotor solidity	σ	0.1
Rotating speed	Ω	40.12 rad/s
Thrust coefficient over solidity	$\frac{C_T}{\sigma}$	0.1
Number of blades	N	4
Lift curve slope	a	5.73
Blade linear twist	θ_{tw}	$-8°$
Mass per unit length	m	6.4636 kg/m
Lock number	γ	6.35
Sectional flap stiffness	$\frac{EI}{m\Omega^2 R^4}$	0.008345

Table 5.2 Non-dimensional rotating frequencies

Mode	$(s=1)$ Frequencies	$(s=2)$ Frequencies	$(s=3)$ Frequencies	$(s=4)$ Frequencies	$(s=5)$ Frequencies
1	3.6816	4.1373	4.7972	5.5850	6.4495
2	22.1810	22.6149	23.3202	24.2733	25.4460
3	61.8417	62.2731	62.9849	63.9667	65.2050
4	121.0509	121.4966	122.2355	123.2614	124.5664
5	200.0115	200.4669	201.2232	202.2767	203.6220

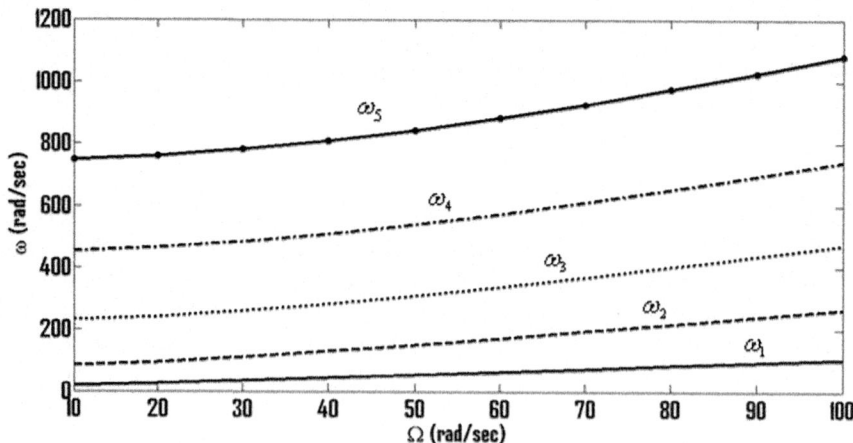

Fig. 5.1 Campbell diagram

5.3 Result 2 (Response of the Rotor Blade with the Uniform Inflow Model, Using Three Different Cases)

Three sets of vehicle control and attitude are considered at three different flight speeds which range from slow speed ($\mu = 0.15$) to high speed ($\mu = 0.35$). The results are shown in Fig. 5.2. The response increases at higher forward speeds.

Case 1 – ($\theta_{1s} = -2.4°$, $\theta_{1c} = 1.9°$, $\theta_0 = 6.75°$, $\theta_{tw} = -8°$, $\alpha_s = 1.2°$, $\mu = 0.15$)

Case 2 – ($\theta_{1s} = -5.4°$, $\theta_{1c} = 1.8°$, $\theta_0 = 8.5°$, $\theta_{tw} = -8°$, $\alpha_s = 3.9°$, $\mu = 0.30$)

Case 3 – ($\theta_{1s} = -6.8°$, $\theta_{1c} = 1.8°$, $\theta_0 = 9.3°$, $\theta_{tw} = -8°$, $\alpha_s = 4.2°$, $\mu = 0.35$)

The results obtained by present formulation are similar to those given in [11].

5.4 Result 3 (Response of the Rotor Blade with the Linear Inflow Model, Using Three Different Cases)

The same three cases are considered with a linear inflow model. Finite element in time is used to obtain the results in Figs. 5.2 and 5.3 with 1 element and 17 nodes.

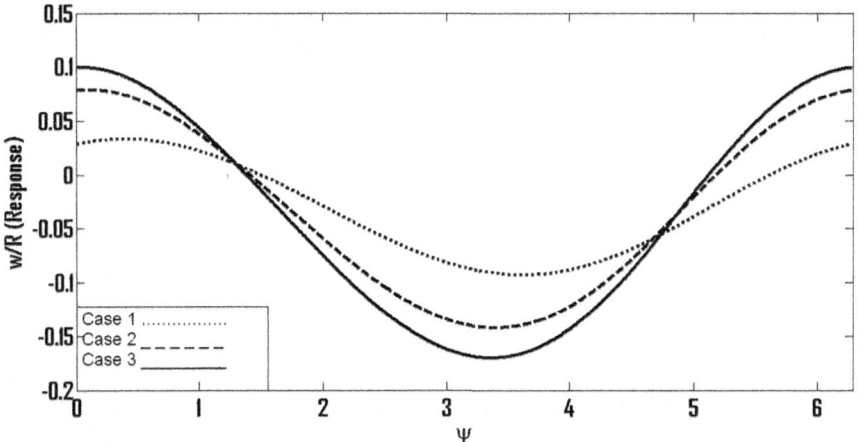

Fig. 5.2 Response with uniform inflow

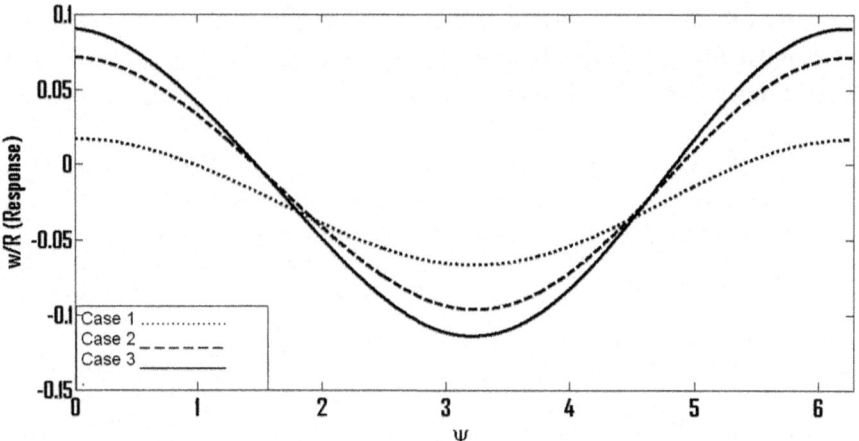

Fig. 5.3 Response with linear inflow

5.5 Result 4 (Stability in Hover Condition)

Advance ratio (μ) goes to zero in hover condition. The stability graph is plotted with the non-dimensional natural frequency (ω/Ω) 1.189 and with varying Lock number. Here, the constant coefficient method discussed in Chap. 4 is used. Initially, as the Lock number is zero, there is no aerodynamic force, and we will get the complex conjugate roots $1.189i$ and $-1.189i$. Our results are similar to [5] (Figs. 5.4 and 5.5).

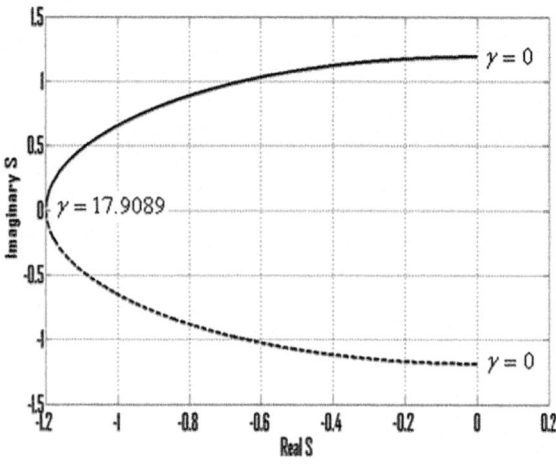

Fig. 5.4 Stability in hover condition with varying Lock number

5.6 Result 5 (Stability in Forward Flight)

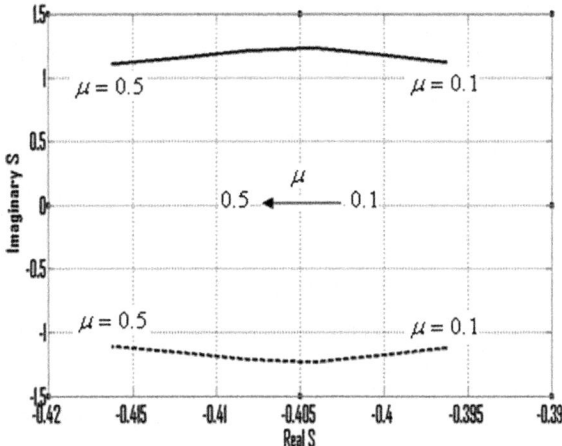

Fig. 5.5 Stability in forward flight condition

5.6 Result 5 (Stability in Forward Flight)

Floquet theory is used to obtain the stability results in forward flight. Non-dimensional frequency (ω/Ω) is taken as 1.189, advance ratio is varied from 0 to 0.5, and the Lock number is kept at 6. Result is similar to [5].

5.7 Summary and Conclusions

This book has outlined an unaddressed area of helicopter dynamics which involves the rotating beam problem. The rotating beam in pure flap bending represents a sound pedagogical model to explain concepts of natural frequency, solution of the governing differential equation, and analysis of the stability of the differential equations. In this book, the rotating beam problem has been systematically developed and solved for helicopter dynamics. The basics of vibration theory are presented, followed by the derivation of the rotating beam governing differential equation including aerodynamic loads. The partial differential equation has spatial and temporal coordinates. The spatial variable is handled through the finite element in space domain, and the time variable is handled using the finite element in time domain. The use of time finite element allows us to enforce the periodic boundary condition for the time domain problem. Once the response of the rotating beam to aerodynamic forcing has been calculated, the stability analysis is performed. The Floquet method of stability calculation is explained for the periodic system. Finally, some solutions generated by the model developed in this book are compared with published literature. The topics developed in this book arms the reader with the necessary tools for understanding and working with sophisticated helicopter rotor dynamics software packages where rotating beams with multiple deformations are typically used.

References

1. Borri, M.: Helicopter rotor dynamics by finite element time approximation. Comput. Math. Appl. **2**(1), part A, 149–160 (1986)
2. Gudla, P.K., Ganguli, R.: Discontinuous Galerkin finite element in time for solving periodic differential equations. Comput. Methods Appl. Mech. Eng. **196**, 682–696 (2006)
3. Zhang, J.: Active-passive hybrid optimization of rotor blades with trailing edge flaps. In: A Thesis in Aerospace Engineering. The Pennsylvania State University (2001)
4. Lim, I.G., Lee, I.: Aeroelastic analysis of bearingless rotors with composite flexbeam in hover and forward flight. In: 16th International Conference on Composite Materials, July 2007
5. Johnson, W.: Helicopter Theory. Princeton University Press, Princeton (1980)
6. Sandilya, K., Ganguli, R., Mani, V.: Non-rotating beams isospectral to a given rotating uniform beam. Int. J. Mech. Sci. **66**, 12–21 (2013)
7. Sheng, G., Fung, T.C., Fan, S.C.: Parametrized formulations of Hamilton's law for numerical solutions of dynamic problems: Part 1 and 2. Time finite element approximation. Comput. Mech. **21**, 441–460 (1998)
8. Friedmann, P.P.: Numerical methods for determining the stability and response of periodoc systems with application to helicopter rotor dynamics and aeroelasticity. Comput. Math. Appl. **12A**, 131–148 (1986)
9. Bauchau, O.A., Nikishkov, Y.G.: An implicit Floquet analysis for rotorcraft stability evolution. J. Am. Helicopter Soc. **46**, 200–209 (2001)
10. Leishman, J.G.: Principle of Helicopter Aerodynamics. Cambridge University Press, New York (2002)
11. Hu, X.Y., Han, J.L., Yu, M.: Nonlinear aeroelastic coupled trim and stability analysis of rotor-fuselage. Appl. Math. Mech. Engl. **31**(2), 237–246 (2010)
12. Lee, I., Jeong, M.S., Yoo, S.J.: Aeroelastic analysis for helicopter rotor blades in hover and forward flight. In: 28th International Congress of the Aeronautical Sciences, 25 Sept 2012
13. Sinha, S.C.: Stability analysis of systems with periodic coefficient: An approximate approach. J. Sound Vib. **64**(4), 515–527 (1979)
14. Meirovitch, L.: Fundamental of vibration. McGraw-Hill Higher Education, New York (2002)
15. Hutton, D.V.: Fundamentals of finite element analysis. McGraw-Hill Higher Education, New York (2005)
16. Rao, S.S.: Mechanical Vibrations. Pearson Education, Upper Saddle River (2004)

© Springer Nature Singapore Pte Ltd. 2018
R. Ganguli and V. Panchore, *The Rotating Beam Problem in Helicopter Dynamics*,
Foundations of Engineering Mechanics, https://doi.org/10.1007/978-981-10-6098-4

Printed by Printforce, the Netherlands